SMART CITY THE NEXT GENERATION
FOCUS SOUTHEAST ASIA

A project by Aedes East International Forum for Contemporary Architecture NPO, in collaboration with the Goethe Institutes in Southeast Asia, with four workshops, an exhibition and a symposium.

Aedes

Content

4 Introduction
Ulla Giesler Curator Aedes, Berlin

10 Preface
Franz Xaver Augustin Goethe Institut, Regional Director Southeast Asia, Australia, New Zealand

12 Smart City Workshop - Grid Structures

14 Collective Studio, Phnom Penh, Cambodia
Tropical Urban Block An Integrated Planning Strategy, Phnom Penh, Cambodia

18 Smart City Workshop - Phnom Penh

20 Hardy Suanto, Architecture Student from Bina Nusantara University, Jakarta, Indonesia
Pluit Reservoir Restoration Biological Environment Restoration Project, Penjaringan, Jakarta, Indonesia

24 Atelier Cosmas Gozali, Jakarta, Indonesia
New Jakarta Green Belt An Urban Planning Vision, Jakarta, Indonesia

28 Research Institute for Humanity and Nature (RIHN), Japan
Alternative Helicopter: Ayun Ayun Kaliku Enhancing the Awareness of Endangered Polluted River, Kampung Cikini Kramat, Central Jakarta, Indonesia

32 Budi Pradono Architects in collaboration with Bina Nusantara University & Tarumanegara University, Jakarta, Indonesia
Fluidscape City Urban Development for Contemporary Reservoir City Project, Penjaringan, North Jakarta, Indonesia

36 Tim Pajus, Medan, Indonesia
Pajak USU Market Revitalization at the North Sumatera University, Sumatra, Indonesia

40 Smart City Workshop - Jakarta

42 ETH Zurich, Future Cities Laboratory (FCL), Singapore
Low-Cost/Low-Tech: Budget Air Travel and the Future of Southeast Asian Cities A Research on Mobility

46 ETH Zurich, Future Cities Laboratory (FCL), Singapore
Fiber Composite Reinforced Concrete A Material Research, Tropical Zone Worldwide

50 W Architects, Singapore
Education Research Center (ERC), NUS A Sustainable Learning Complex, Singapore

54 Eleena Jamil Architect, Kuala Lumpur, Malaysia
HOME Units Low Income Housing for the Elderly, Singapore

58 Arkomjogja, Yogyakarta, Indonesia
Bamboo Bridge Community-Driven Upgrading of an Urban Informal Settlement, Davao City, The Philippines

60 Manon Otto, Montreal, Canada, Student at Lund University, Sweden
Warning – Pollen Landing Shelter Housing, Manila, The Philippines

64 Palafox Associates, Makati City, Metro Manila, The Philippines
San Juan City Plan Planning Towards a Future of Green Consciousness, Manila, The Philippines

68 Smart City Workshop - Manila

70 AND Development Co., Ltd, Bangkok, Thailand
Awarehouse Upgrading Factories and Employee Housing, Bangkok Metropolitan Area, Thailand

74 **Shma Company Limited, Bangkok, Thailand**
Integrated Park and Streetscape – Chulalongkorn University Centennial Park Resilient Park for an Overcrowded City, Bangkok, Thailand

78 **TYIN tegnestue, Trondheim, Norway**
Klong Toey Community Lantern Public Spaces Built with Students and the Community, Bangkok, Thailand

80 **Franken Architekten, Frankfurt am Main, Germany**
U-Silk City Urban Design of 13 Residential High-Rise Towers with Mixed-Use Podium, Hanoi, Vietnam

84 **CAt (Coelacanth and Associates), Tokyo, Japan**
Ho Chi Minh City University of Architecture New Campus for an Architecture University, Ho Chi Minh City, Vietnam

88 **RT+Q Architects, Singapore**
D7 & D6 Building Rethinking Urbanism and the Office Prototype, Kuala Lumpur, Malaysia

92 Acknowledgements

96 Imprint

Einleitung

„Smart City: The Next Generation" spielt mit dem doppeldeutigen Begriff der nächsten Generation: Ein Upgrade der Stadt und ihre zukünftigen Bewohner. Der regionale Fokus liegt dabei auf dem Teil der Welt, in der die globale Zukunft maßgeblich bestimmt werden wird: Süd-Ost-Asien.

In drei vorbereitenden Workshops in Phnom Penh, Jakarta und Manila, initiiert gemeinsam mit dem Goethe-Institut, versuchten wir den Blick der jungen Generation einzufangen: Wie sehen sie ihre Städte? Wie wird die Zukunft gestaltet? Mit einem Call for Proposals luden wir darüber hinaus Architekten, Planer, Universitäten, Initiativen und Künstler ein, ihre Entwürfe für die morgigen Städte mit uns zu teilen. Aus über 90 Einsendungen haben wir knapp 40 ausgewählt: Visionen, Masterpläne, Community-Aktionen, Gebäude und Infrastrukturprojekte, die die Städte intelligenter, wirkungsvoller und – vor allem – lebenswerter machen. Sie zeigen oder fordern Lösungen zu vielfältigen Problemlagen, dem Umgang mit der knappen Ressource Wasser, für neue Verkehrskonzepte oder dem Einsatz regionaler Materialen. Sie stellen provokante Entwürfe in den Raum: Mobile Behausungen für unterprivilegierte Menschen in Singapur, urbane Gärten und Wohnungen für Arbeiter der Fabriken Bangkoks, sie „verpflanzen" Slumbewohner aus den Uferzonen verdreckter Flüsse auf bestehende Häuser und bieten den Hausbesitzern dafür Regenwassergewinnung an oder zeigen Studenten, dass Lernen im Freien mit natürlicher Belüftung selbst in tropischen Regionen ohne Verzicht auf Komfort möglich ist. Dabei stehen realisierte Projekte arrivierter Architekturbüros gleichberechtigt neben Ideen junger Absolventen, zukünftige Planungen lokaler wie europäischer Städtebauer neben Recherchen und Studien renommierter Institutionen, kreative Interventionen junger Designer neben sozialen Initiativen.

Drei Fragen verbinden alle Projekte und ziehen den roten Faden durch Ausstellung und Katalog:
WIE macht Ihr Projekt die Stadt „smarter"?
WAS sind die spezifischen Herausforderungen in Ihrer Stadt?
WELCHE neuen Verhaltensweisen bewirkt Ihr Projekt?

Im Wintersemester 2012/2013 entwickelten Design-Studenten in Düsseldorf Ideen zur Präsentationsform der süd-ost-asiatischen Workshops. Aus sechs Ausstellungsarchitekturen entschieden wir uns für ein aus Klettband gespanntes Netzwerk, „Smart Grid", das in der Ausstellung die Städte Phnom Penh, Jakarta und Manila verbindet. Im Sommersemester stand die herausfordernde 3-dimensionale Umsetzung des Grids, sowie die Verschränkung mit den von mir gemeinsam mit Christine Meierhofer entwickelten analogen „Loops" an, die das umfassende Material der ausgewählten Projekte zeigen. Auch der Ausstellungsbesucher darf auf einer spielerischen Ebene direkt eingreifen und sich die Projekte anhand der Fragen vergleichbar machen, indem er die „Loops" nach unten oder oben scrollt.

Neben der Ausstellung thematisiert ein zweitägiges Symposium die Frage nach „smarten Verhalten" und setzt damit – wie beabsichtigt – den Menschen in den Mittelpunkt des Projekts. Im anschließendem Workshop am Aedes Network Campus Berlin forschen die eingeladenen Young Professionals aus Süd-Ost-Asien gemeinsam mit europäischen Kollegen.

Das weitgefächerte „Next Generation"-Projekt setzt damit die Reihe der Aedes-Ausstellungen fort, die im Rahmen der Asien-Pazifik-Wochen Berlin entstanden und die weit über die Architektur hinaus urbane Gefüge beleuchten und die Zusammenschau regionaler Ansätze ermöglichen. Diese Ausstellungen fanden große Rückstrahlkraft nach Asien. So wurde die Ausstellung „Water – Curse or Blessing?!" 2012 als erste internationale Architekturausstellung in Phnom Penh gezeigt. Das Goethe-Institut in Kuala Lumpur lud mich Ende 2012 ein, einzelne APW-Projekte auf einer Konferenz mit malaiischen Studenten zu diskutieren.

Mit „Smart City: The Next Generation" wurde im Besonderen möglich, wofür Aedes steht: Der Förderung junger Talente, Öffentlichkeit zu schaffen für urbane Fragestellungen und nachhaltige Netzwerke zu bilden. So endet das Projekt auch nicht in Berlin, sondern wird 2014 unter der Federführung des Goethe-Instituts durch Süd-Ost-Asien touren.

Möglich wäre das nicht ohne die großzügige Unterstützung der Stiftung Deutsche Klassenlotterie Berlin und des Goethe-Instituts, in Person des Regionalleiters für Süd-Ost-Asien Franz Xaver Augustin, die den Großteil der Finanzierung übernahmen. Ihnen sei besonders gedankt.
Danken möchte ich auch Volker Schlegel, Präsident des Asien-Pazifik-Forum Berlin, der mich ermutigte, dieses Projekt in Angriff zu nehmen und – auch das muss mal gesagt sein – in den Dschungel der Finanzierung einzusteigen; allen Senatsabteilungen, die meinen Einsatz zur Brückenbildung zwischen Berlin und Asien seit Jahren mit ihrer Wertschätzung unterstützen; den Kooperationspartnern von Aedes, Zumtobel, Busch Jaeger, Carpet Concept und besonders Axor Hansgrohe, die das noch fehlende „Brückengeländer" spendiert haben sowie Transsolar, der ETH Zürich/Future City Lab und dem Auswärtigen Amt, das Flüge für Studenten aus Phnom Penh und Jakarta übernommen hat.

Und vor allem all denen, die wunderbare Projekte eingereicht oder für uns in den Workshops entwickelt haben. Genannt seien hier die Leiter der Workshops: Shelby Doyle mit Unterstützung des Meta-House in Phnom Penh; Elisa Sustanudjaja/Marco Kusumawijaya von Rujak gemeinsam mit Dietmar Leyk und dem Goethe-Institut in Jakarta; ebenso Dietmar Leyk mit der Gruppe um Joey Yupangco mit Unterstützung von Luisa Zaide, Goethe-Institut Manila.

Und natürlich Kristin Feireiss und Hans-Jürgen Commerell, die das Fundament bilden.

Ulla Giesler
Initiatorin, Kuratorin und Projektmanagerin

Introduction

"Smart City: The Next Generation" plays on the ambiguity of the term "next generation". An upgrade of the city and its future inhabitants. The regional focus lies on the part of the world that will be defining our global future to a great extent: Southeast Asia.

In three workshops in Phnom Penh, Jakarta und Manila, organized together with the Goethe Institut, we tried to capture the perspectives of the young generation: How do they see their cities? How will the future be designed? With a "Call for Proposals" we also invited architects, urban planners, universities, initiatives and artists to share their designs for the city of tomorrow with us. From over 90 entries we chose 40: visions, master plans, community actions, buildings and infrastructure projects that make the cities more intelligent, more effective and above all raise the quality of life there. They show or demand solutions to a wide range of problematic situations, the handling of scarce resources like water, new transportation concepts or the use of local materials. They pose provocative questions: mobile housing for underprivileged people in Singapore, urban gardens and homes for factory workers in Bangkok, they "transplant" slum dwellers from the banks of polluted rivers onto existing houses and offer the house owners rainwater supply in return, or show students that studying in the open air with natural air conditioning, even in tropical regions, is possible without having to compromise on comfort. Presented are realized projects from renowned architects side by side with new ideas by young graduates, future plannings of local and European urban planners next to research papers and studies by renowned institutions, creative interventions by young designers next to social initiatives.

Three questions connect the projects and serve as the thread through exhibition and catalogue:
HOW does your project "smarten" up your city?
WHY does your city need your project?
WHAT are the new behaviours encouraged by your project?

In the winter semester 2012/2013, design students in Düsseldorf developed ideas for the presentation of the Southeast Asian workshops. Out of six exhibition designs we chose a network of Velcro, titled "Smart Grid", that interconnects the cities of Phnom Penh, Jakarta and Manila in the exhibition. The summer semester was spent with the challenging three-dimensional realisation of the grid and its linking with analogue "Loops", that were developed by Christine Meierhofer and me to present the extensive material of the selected projects. The visitor is invited to interact on a playful level, and make comparisons between the projects by scrolling the "Loops" up or down.

In addition a two-day symposium takes a closer look at the theme of "smart behaviour" and – as intended – places people firmly at centre stage of the project. In the subsequent workshop at Aedes Network Campus Berlin the young professionals who have been invited here from Southeast Asia will continue their research together with European colleagues.

The wide-ranging "Next Generation" project is a continuation of the series of Aedes exhibitions that were created in the framework of the Asia Pacific Weeks Berlin (APW) and whose scope reaches way beyond that of architecture, illuminating the urban fabric and enabling a comparison of different regional approaches. The exhibitions to date had a great impact back in Asia too. The exhibition "Water – Curse or Blessing?!" was shown in Phnom Penh in 2012 as the first international architectural exhibition. The Goethe Institut in Kuala Lumpur invited me to discuss individual APW projects at a conference with Malayan students at the end of 2012.

With "Smart City: The Next Generation" the features for which Aedes is so well known are very much present: the advancement of young talent, creating public discourse on the issues of urban concepts and the creation of long-term networks. In this spirit, the project will not end in Berlin, but will continue touring through Southeast Asia in 2014 under the auspices of the Goethe Institut.

This would all not have been possible without the generous support of the Stiftung Deutsche Klassenlotterie Berlin and the Goethe Institut, represented by the regional director for Southeast Asia, Franz Xaver Augustin, who took on a large part of the financing. A special thanks to them. I would also like to thank Volker Schlegel, President of the Asia Pacific Forum Berlin, who encouraged me to develop this project and – this aspect should also be mentioned – to find my way through the jungle of funding; all senate departments who have supported me and appreciated my commitment to building bridges between Berlin and Asia over the years; Aedes' cooperation partners Zumtobel, Busch Jaeger, Carpet Concept and in particular Axor Hansgrohe, who sponsored the missing "bridge railings" metaphorically speaking, as well as Transsolar, ETH Zürich/Future City Lab and the Ministry for Foreign Affairs who took on the financing of flights for students from Phnom Penh and Jakarta.

And above all, thanks to those who entered their wonderful projects or developed the workshops for us. The workshop heads were: Shelby Doyle with support from Meta-House in Phnom Penh; Elisa Sustanudjaja/Marco Kusumawijaya from Rujak together with Dietmar Leyk and the Goethe Institut in Jakarta; as well as Dietmar Leyk with the group around Joey Yupangco with support from Luisa Zaide, the Goethe Institut in Manila.

And of course Kristin Feireiss and Hans-Jürgen Commerell, who are the foundation of all this.

Ulla Giesler
Initiator, curator and project manager

Exhibition at the Aedes Architecture Forum in Berlin, May 2013

Vorwort

SMART CITY – Eine Kooperation von Aedes East e.V. und dem Goethe-Institut

Seit Schulzeiten sind uns die exotischen Namen und die bizarre Gestalt der Landzungen und Inseln zwischen Australien und dem asiatischen Festland geläufig. Dieses auffällige Puzzle aus Land und Meer umfasst die Länder Süd-Ost-Asiens, die sich vor mehr als 50 Jahren zur Gemeinschaft der ASEAN-Staaten zusammengeschlossen haben. Seit Jahrhunderten bilden sie eine Welt des Durchgangs, des intensiven Austauschs und einzigartiger kultureller Vielfalt. Fast 600 Millionen Menschen wohnen hier, was etwa der Hälfte der Einwohner Indiens entspricht – die Mehrheit in Städten, von denen viele strategisch günstig an den großen Routen des Welthandels liegen. Ihre maritime Lage brachte es mit sich, dass viele dieser Zentren den europäischen Kolonialmächten als Stützpunkte dienten und als solche ihren ersten großen urbanen Aufschwung erlebten. Gemeinsam ist ihnen weiter, dass sie in den Jahrzehnten nach der Unabhängigkeit mit frappierender Geschwindigkeit wuchsen und ihre Einwohnerzahl in kürzester Zeit vervielfachten. Dabei entstanden Metropolen, die in ihrer urbanen Organisation und Infrastruktur, ihrer Architektur und ihren Angeboten an Lebensqualität nicht unterschiedlicher sein könnten. Neben dem geradezu ideal organisierten Stadtkunstwerk Singapur stehen auf der anderen Seite der Skala wild wuchernde Megastädte von der Art Jakartas oder Manilas, die in oft verstörender Missachtung ökologischer Grundprinzipien alle Annehmlichkeiten städtischen Lebens zu zerstören drohen. Doch gerade hier, wo die gestalterische Intervention der öffentlichen Hand so oft zu wünschen übrig lässt, gibt es eine wachsende Zahl junger Architekten, engagierter Stadtplaner und ökologischer Aktivisten, welche die Herausforderungen ihrer urbanen Umgebung als Ansporn begreifen und mit umso entschlossenerer Findigkeit und kreativer Frische nach neuen Lösungen für die geschundenen städtischen Envirements suchen.

Dass diese oft gegen viele Widerstände vertretenen Ansätze einer jungen Generation von Architekten und Urbanisten in den Blick genommen werden und dabei nicht nur die bekannten Beispiele der beiden asiatischen Großmächte China und Indien aufscheinen, sondern Modelle aus den gewöhnlich weniger beachteten Metropolen Süd-Ost-Asiens in den Mittelpunkt rücken, ist ein großes Verdienst des Smart City Projekts. Der Einladung seiner Initiatorin Ulla Giesler und des Aedes East e.V. sind die Goethe-Institute in Süd-Ost-Asien mit großer Freude gefolgt und haben ihre bewährten Kontakte zu jenen Gruppen junger Stadterneurer gern zugänglich gemacht, die nun in dieser Ausstellung mit ihren Entwürfen zu sehen sind. Mein Dank gilt dem Team von Ulla Giesler und Aedes, Shelby Doyle für die Leitung des Workshops in Phnom Penh, Elisa Sutanudjaja für die Organisation des Workshops in Jakarta sowie Dietmar Leyk als den inspirierenden Kopf der Workshops in Jakarta und Manila, meinen Kollegen in Manila, Jakarta, Phnom Penh und Kuala Lumpur und all den stadtbegeisterten Teilnehmern an dieser Ausstellung.

Jakarta, 20. Mai 2013
Franz Xaver Augustin
Regionalleiter der Goethe-Institute in Süd-Ost-Asien

Preface

SMART CITY – a joint project of Aedes East NPO and the Goethe Institut

The exotic names and bizarre shapes of the tongues of land and islands that lie between Australia and the Asian mainland are familiar to us all from our schooldays. This striking patchwork of land and sea encompasses the countries of Southeast Asia, which almost 50 years ago joined together to form the Association of Southeast Asian Nations, or ASEAN for short. For centuries this region has been a place of transit, intensive exchange and unique cultural diversity. Nearly 600 million people live here, equivalent to roughly half the population of India – the majority of them in cities, many of which are strategically located along the major global trading routes. Thanks to their maritime situation, many of these centres were used as staging posts by the European colonial powers and as such experienced their first major urban boom. Another thing they have in common is that they grew at tremendous speed following independence, their populations multiplying many times over within a very short space of time. This gave rise to metropolitan centres whose urban organization and infrastructure, architecture and lifestyles could hardly be more different. In stark contrast to the urban artwork that is Singapore, a city of almost utopian organization, we have at the other end of the scale megacities such as Jakarta and Manila that have been allowed to grow unchecked, urban sprawls created with an alarming disregard for basic ecological principles that threaten to destroy all the conveniences of modern city life. Nonetheless, it is precisely in such cities, where the public sector so often fails to intervene sufficiently in matters of urban development, that we find a growing number of young architects, dedicated city planners and ecological activists who view the challenges posed by their urban landscapes as stimulating and who are all the more resolved to employ their resourcefulness and creative vigour in a quest for new solutions to the damaged urban environments.

One of the great merits of the Smart City project is that it explores the work and ideas of a young generation of architects and city planners – whose approaches frequently meet with considerable resistance – and not only showcases the examples that we are all familiar with from China and India, Asia's two major powers, but also takes models from Southeast Asian cities that tend to be paid less attention and places them centre stage. The Southeast Asian branches of the Goethe Institut were delighted to accept the invitation from the project's initiator Ulla Giesler and Aedes East NPO and to grant access to their established contacts among those groups of young urban renewers whose designs now feature in this exhibition. I would like to thank Ulla Giesler's team at Aedes, Shelby Doyle for running the workshop in Phnom Penh and Elisa Sutanudjaja in Jakarta, as well as Dietmar Leyk for his great inspirational role in the workshops in Jakarta and Manila, my colleagues in Manila, Jakarta, Phnom Penh and Kuala Lumpur, and everyone who participated in this exhibition with such urban enthusiasm.

Jakarta, 20 May 2013
Franz Xaver Augustin
Regional Director Goethe Institut Southeast Asia

Smart City Workshop - Grid Structures

Düsseldorf University of Applied Sciences / Faculty of Design, Germany
Winter semester 2012/13 + summer semester 2013

"The grid is the fundamental organizing structure that defines the work in terms of formal properties as well as the process of creating. It is the single constant through the process - everything else is open to varying degrees of manipulation and subjectivity."[1] Willam Betts

Within the scope of the design seminar Grid Structures a group of 23 students from Düsseldorf University of Applied Sciences, Faculty of Design, developed spatial concepts based on the notion of grid and network structures: open spatial interfaces for the exhibition "Smart City: The Next Generation" at the Aedes Architecture Forum in Berlin.

By instrumentalising basic spatial elements – points, lines and volumes in space, a great variety of the students' works resulted in grid-like and network constructs, walk-in and "usable" spatial geometries and relational mesh organisations that were in their materiality fixed, elastic, flexible, rigid or movable. All of those developed structures became "carriers" for a vast amount of architectural and artistic projects, dealing with the question of a "smart city" that are now part of the exhibition. All of the developed spatial grids had in common the fact that they created non-architecture-like open spaces without roofs, walls or doors, defining rather open zones than boundaries or closed spaces – enabling interaction and change with its users.

It was specifcally during the 60s, where in architecture many examples were to be found in which architects were investigating grid-like spatial organisations to realize social and infrastructural relations of space. Buckminster Fuller and later Yona Friedman or Archigram developed spatial models and geometries where architecture merely provided a framework in which the inhabitants and users might construct their own spaces according to their own needs. Spatial networks were considered not to be merely visual, but instead became the Modus Operandi themselves for their inhabitants. Those networks became reacting systems triggering interaction with their users. Today in contemporary art there is for instance the architect and artist Tomás Saraceno who investigates the notion of the network on a constructive and atmospheric level. His spatial geometries articulate themselves physically in real space, instrumentalising specific materialities and techniques of construction.

Project Supervisors Prof. Gabi Schillig, Räumlich-Plastische Gestaltung, Studio for Spatial Design, Düsseldorf University of Applied Sciences / Faculty of Design - in cooperation with Ulla Giesler, Aedes

Students Seminar Grid Structures
Stephanie Butzen, Yiqing Cai, Teresa Christ, Marcel Czeczinski, Rouven Dürre, Tabea Faß, Wibke Gocht, Iris Hamers, Nadine Hofmann, Xiao Fan Jin, Andrea Krämer, Adalbert Kuzia, Marie Märgner, Bea Meder, Nadine Nebel, Niklas Reiners, Carolin Reisensohn, Jonas Schneider, Jenni Stark, Jana Stenzel, Luis Torres, Janina Ungemach, Sina Wassermann

Selected Exhibition Structure "Klettwerk"
Design and Concept Rouven Dürre, Marie Märgner, Adalbert Kuzia, Tabea Faß
Realisation Adalbert Kuzia, Nadine Hofmann, Marie Christine Keppler, Nadine Nebel, Jonas Schneider, Anna Wibbeke, in cooperation with Michael Swottke, Model Building Workshop Düsseldorf University of Applied Sciences

The selected exhibition structure "Klettwerk" that now has been realized by a group of students within the framework of the Smart City exhibition, consists of spatial lines that are made out of Velcro tape, tensed up into Aedes' gallery space. The interlocking and joining of those black linear ribbons provides a levitating open grid-and city-like structure, an open space that is both continuous and multi-directional. The line's materiality (hook-and-loop band) and its geometrical manifestation in space offers opportunities for visitors to perform, to participate, to change. The construct becomes a mediator between the gallery space, the content of the exhibition and its visitors, changing their behaviour and therefore experience in space. By moving and re-arranging Smart City projects, leaving comments on existing concepts, the linear exhibition structure is subject to constant re-organization and becomes an open, flexible spatial system that "behaves" itself in a performative and adaptive manner.

The realized spatial grid anticipates the interaction of its spectators with their environment. This performative process breaks the boundary between traditionally divided units and enables the communication between space, content, spectator and results in the dissolution of spatial boundaries which orginally separated both of them. Space realizes itself in the action and imagination of the spectator who has become an active participant and user, without whose physical participation the concept would not be entirely realized. The space itself possesses an inventive strength and emerges as something that is never definite or complete. The organising grid structure offers an open system, an opportunity for communication that is, despite organizing the gallery space into open zones (and not into an architecture of rigid walls) open enough for transformation and change. Like a "smart city" it offers options for action, operation and change for its inhabitants and users.

Gabi Schillig
Berlin / Düsseldorf, Spring 2013

1 in: Rasterfahndung, Kunstmuseum Stuttgart / edited by Ulrike Groos, Simone Schimpf / Wienand Verlag Köln, 2012

Tropical Urban Block
An Integrated Planning Strategy, Phnom Penh, Cambodia

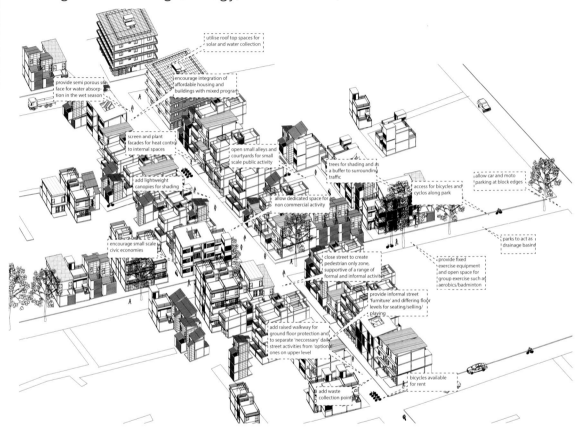

Phnom Penh, the capital of Cambodia, is rapidly developing. At the moment there is seemingly no reference model for public urban space that is sensitive to the local context. Via a series of small scale interventions, this project proposes to reorganize a city block in order to create an alternative model that:
- is inclusive and participatory
- is responsive to climatic challenges
- encourages more sustainable behaviour
- is scalable

Office Collective Studio, Phnom Penh
Location Phnom Penh, Cambodia
Design period End 2012 / 2013
Realisation period Unbuilt, in mid 2013 Collective Studio will present the project to different groups of citizens
Concept by Giacomo Butte
Team Sarah Bland, Eva Esposto Lloyd
Website www.collectivestudio.cc

HOW?

ACTIVATE BUILDINGS

- rooftop activities
- extend balconies for solar shading
- second street level
- extend ground level to create colonnade

1 creates urban space which is inclusive and participatory
provides non-privatised space open to a wide mix of users and activities and encourages the mix of these
requires active collaboration of citizens, policy makers and designers to develop, implement and sustain

2 creates urban space which is responsive to climatic challenges
provides protection from, and makes use of, tropical sun and rain
manages wet season flooding

3 creates urban space that encourages more sustainable behaviour
useable outdoor space protected from the elements as opposed to internal air conditioned environments
active space for pedestrians only to limit the use of vehicles

4 proposes an alternative model of urbanization to the current, which is repeatable
encouraging policies and city making which promote public space accessible to all

PUBLIC SPORT AREAS

access to exercise equipment and non-commercial activity areas

STREET HIERARCHY

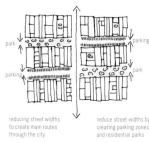
- reducing street widths to create main routes through the city
- reduce street widths by creating parking zones and residential parks

ACTIVATE ALLEYWAYS

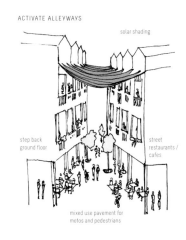
- solar shading
- step back ground floor
- street restaurants / cafes
- mixed use pavement for motos and pedestrians

RAISED WALKWAYS
raising pavements for flood prevention

- solar shading
- informal seating
- rainwater drainage
- stepped seating

ENVIRONMENTAL DESIGN

- solar panels
- rainwater collection

GREEN BASIN
rainwater run off deposited into water basins in the street

new gardens created on residential streets

ACTIVITY PODIUM
- informal central street podiums
- traffic calming measure on residential streets
- shaded canopy
- raised podium for flood prevention

RUBBISH COLLECTION POINTS

recycling points for all residential blocks

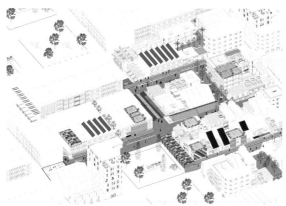

Clustered within the area of a block, a series of small interventions modifies the existing city structure.

The typical urban fabric of Phnom Penh. In the foreground the open sewer that runs north to south along street 105 with a view of Boeung Trabek market and further east to the Mekong River on the horizon.

WHY?

1 growing popularity of privatised, internal space
rise of shopping malls and air conditioned coffee shops

2 non pedestrian-friendly city
no public transport system
heavy reliance on vehicles
sidewalk space as parking space

3 poor drainage management
insufficient infrastructure to cope with wet season flooding
increasing strain from lake infilling, population growth and climate change

4 poor waste management
insufficient systems for trash and sewage

5 lack of transport infrastructure
difficult to promote pedestrian zones and vehicle minimisation

6 governance encouraging non inclusive attitudes
history of urban poor evictions, in favour of private development
no provision for affordable housing and little support for the marginalised
resistance to public expression

7 weak regulatory framework
master planning and building codes loosely implemented
land ownership unclear
land value and speculation is increasing

8 rapidity of development
'development' focuses on short term, private, economically-driven projects

The current urban structure of the city does not include any particular strategy for climate mitigation.

Vegetation is used for climate mitigation. Road structure is designed to improve water absorption.

Phnom Penh sidewalks are usually occupied by cars, motos and shops.

Streets, turned into pedestrian areas, are equipped with facilities for public activities and performances. Some buildings are turned into public facilities.

Phnom Penh does not have a pedestrian area beside the river side.

1 creating a vehicle free pedestrian zone with protection from the elements
promotes a change in thinking for urban planning and zoning governance
encourages walking and outdoor activity, to minimise environmental impact and maximise interaction

2 creating generic urban components
promotes varied interactions and inclusive behaviours by supporting a wide variety of activities

3 creating open, shaded park areas and designated exercise equipment
encourages healthy behaviour

4 efficiently using natural resources, sun and rain
encourages systems that take advantage of the tropical climate rather than working against it

5 creating elements that protect from sun and rain
encourages the use of public space throughout the day, rather than just morning/evening

WHAT?

Existing buildings do not have an external canopy. Often house owners add plastic canopy to increase the shaded area.

Raised central podium for slower activity such as selling, eating, gathering and flood proofing.

Raised walkway for protection from elements and stimulation of upper levels. This provides shading, increases public areas and connection. Vegetation is used especially as a climate mitigation element.

Smart City Workshop - Phnom Penh
November 3 - December 2, 2012, at MetaHouse, Phnom Penh, Cambodia

In November/December 2012 an Aedes Workshop, lead by Shelby E. Doyle, took place was at MetaHouse, the German Cambodian Cultural Center in Phnom Penh under the title „Phnom Penh Smart City" in the framework oft the exhibition project „Smart City: The Next Generation"; Focus Southeast-Asia.

A group of 30 students from several Cambodian universities participated: Royal University of Fine Arts, Limkokwing University of Creative Technology, Cambodian Mekong University, Norton University and Pannasastra University of Cambodia. They discussed the current challenges facing Phnom Penh and developed design strategies for its future. The aim was to generate projects, which will be displayed at the Aedes Architectural Forum in Berlin.

The initial plan was to limit the scope of the projects and discussions to the current environmental challenges facing the city, particularly the challenges of urban flooding. For workshop coordinator Shelby E. Doyle this is one of the most pressing challenges of Phnom Penh. When she came to Cambodia to research urban development in Phnom Penh, she documented the relationships between water, architecture, and infrastructure in the city. Her objective was to record the architectural and urban conditions sustained by and subject to the cyclical floods of the city's rivers, to describe the challenges faced by Phnom Penh as it rapidly urbanizes in a flood plain, and to explore the nature and agency of design in relation to these topics.

However it quickly became apparent that the challenges Ms. Doyle documented for Phnom Penh were different from those the students perceived for their city. Therefore, she encouraged the students to propose projects that inspired them and out of this work came several projects focused on 'greening' the city, producing walkable sidewalks, and public space.

Project Supervisor Shelby E. Doyle, Fulbright Researcher

Students El Vanthat San, Virak Hin, Pheaksovann Hin, Chenda Va, Chanmolin Kun, Kao Sokly, Thoeu Piranuch, SOK Muygech, Charlotte Lemahieu, Vicheth Damardi, Sreyroth Nun Moni, Vanna Pheaktravisal, Virak Roeun, Vannita Som, Lorenzo Maria Martini, Sotahrin Saran

The exhibition in Berlin shows three projects from this workshop
Mobility by Sotharin SARAN
Rethinking Urban Housing by Muygeck SOK
Busstop by Vannita SOM

Informal discussion of these issues were supplemented by formal presentations:

Week 1, November 3, 2012 Introduction "Phnom Penh Smart City"

Introduction Shelby E. Doyle, Fulbright Researcher / Workshop Coordinator, www.cityofwater.wordpress.com
Open source mapping Nora Lindstrom, Urban Voice Phnom Penh, www.urbanvoice.net
Student conceptual work for the Arch Prix Asia Competition in Phnom Penh www.archiprixsea.com
Discussion of what defines a 'Smart City"

Week 2, November 10 2012 Low-tech 'Smart' Solutions for Phnom Penh

Urban Wetland Pissoir Wetlands-Work! team, http://www.facebook.com/pages/Urban-Wetland-Pissoir/288058214636148
Bus stop - design and public transport system for Phnom Penh challenges of traffic and public, Vannita SOM, http://urbanlabphnompenh.wordpress.com/2012/10/17/bus-stop-the-urban-lab/

Week 3, December 2, 2012 Master Planning for the Future of Phnom Penh

Phnom Penh Master Plan Phanin CHEAM, Vice-chief of urban planning office, Municipality of Phnom Penh and Lecturer at Norton and Mekong University

Additionally, their work addressed the challenges of a city without a mass transit system. Traffic during rush hours brings the city nearly to a halt and frequent traffic accidents, injuries and death remain an ongoing concern for a rapidly development city. As life-long residents of Phnom Penh the students identified these as the city's most urgent design and infrastructure issues.

Pluit Reservoir Restoration
Biological Environment Restoration Project, Penjaringan, Jakarta, Indonesia

The Biological Environment Restoration Project is a concept of restoring or rejuvenating a slum area around the Pluit Reservoir, a water storage reservoir that prevents flooding during the rainy season in Pejaringan, North of Jakarta. The reservoir can be harmful to its surrounding community and the goal is to re-establish a balanced and harmonic life within this environment. The biological concept is intended to use the technology of nature to restore something that has been damaged. Over time, people have started to set up illegal housing on the edge of the Pluit Reservoir. Problems like pollution of the aquatic ecosystem followed by the reduction of fish stock emerge from siltation, uncontrolled wild plants and foremost the disposal of industrial waste into the reservoir or into the river that ends in the reservoir.

The project contains different concepts:
A double filtration system. With local bamboo and construction techniques, layers of plantation are created at the edge of the reservoir to filter the water to support aquaponic production with floating vegetables in the last layer. The amount of loose garbage, which passes the existing filtration at the entrance is a major contamination source of the reservoir. The biological filtration technique is using the ability of nature and bio-organisms such as the reed plant, water hyacinth and Lemna minor to clean the water in the reservoir.

Symbiosis House is new housing, designed with local bamboo on the top of existing legal houses, so-called host houses. With 2 levels its design adapts the way of life in traditional Indonesian houses with a day time communal area at the first level and a sleeping area at the second level. The roof of the Symbiosis House includes a rain harvesting system. With the creation of vacant land that has been cleaned from existing slum houses, public toilets, waste and energy treatments, rain harvesting systems, open space and community areas are established.

Office Hardy Suanto, Architecture Student from Bina Nusantara University, Jakarta, Indonesia
Location Penjaringan, Jakarta, Indonesia
Design period 16 Aug 2012 - 1 Nov 2012
Construction/Realisation period unbuilt
Concept by Hardy Suanto
Team Tri Apriliana
Tutors/Lecturers: Albertus Galih Prawata and Michael Isnaeni Djmantoro
Website architecture.binus.ac.id/2012/12/20/juara-2-architectural-design-archiprix-sea-2012

- BIOLOGICAL FILTRATION - REED
- BIOLOGICAL FILTRATION - WATER HYACINTH
- BIOLOGICAL FILTRATION - WATER SPINACH
- AQUAPONIC PRODUCTION SYSTEM
- EXISTING SLUM HOUSES - BUILT ON ILLEGAL LOT
- EXISTING HOUSES AS HOST FOR NEW HOUSES
- WASTE AND ENERGY THREATMENT
- COMMUNAL SPACE AND RAIN WATER HARVESTING SYSTEM
- LOSS GARBAGE FILTRATION
- EXISTING GARBAGE DISPOSAL

Master Plan Zoning The whole concept is to change the problem in the Pluit Reservoir into potential that can fulfill the people's needs by other point of view and a new system. The potential for development in this area uses the concept of symbiosis and biomimikri as an approach to imitate how nature works in a life cycle.

The project creates a **self-sustaining region** in terms of food and clean water production, waste treatment, housing treatment, community development, economical sustainability and environmental sustainability for the reservoir and surrounding environment. Toxic water is filtered and used for existing plantations and for **a new aquaponic production zone, illegal housing is solved** with the help of Symbiosis Housing, clean water and energy is provided for the daily consumption of the community, **waste management is initiated,** the **natural environment is protected, the health of the environment and the people is restored.**

HOW?

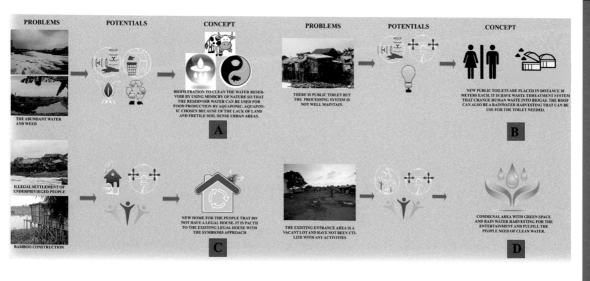

ORGANIC AND ANORGANIC FILTRATION SYSTEM

ORGANIC AND ANORGANIC FILTRATION SYSTEM IS USED TO FILTER LOOSE GARBAGE THAT CAN'T BE FILTERED BY MAIN FILTRATION MACHINE. IN THE EXISTING CONDITION, SOME BAMBOO STICKS HAVE BEEN USED TO HOLD LOOSE GARBAGE PREVENTING IT TO ENTER THE RESERVOIR. THE WORKERS COLLECT THE LOOSE GARBAGE MANUALLY BY CRAFT, THEN MOVE IT TO THE EMPTY LOT IN SITE. THIS CONDITION CAUSE THE POTENTIAL EMPTY LOT LOSS IT CAPABILITY TO SERVE AS OPEN SPACE AND COMMUNAL SPACE.

THIS FILTRATION USES THE SAME METHOD AND MATERIAL AS THE EXISTING GARBAGE HOLDER BUT NEW SYSTEM IS APPLYED TO MAXIMIZE IT'S OPERATION. THIS NEW FILTRATION HAS 3 TIMES FILTRATION LAYERS. WORKERS CAN COLLECT THE GARBAGE EASILY WITHOUT RAFT AS BEFORE BECAUSE A WALKWAY IS INSTALLED TO HELP THE JOB.

AFTER THE COLLECTING JOB IS COMPLETE, GARBAGE IS PLACED INTO A BAMBOO BOX THAT LIFT UP THROUGH PULLEY SYSTEM AND TRANSFER THE WASTE TEMPORARY DISPOSSAL.

OXYGEN FROM ALGAE PLANTATION

UNLIKE OTHER CROPS, ALGAE DO NOT REQUIRE CLEAN WATER IN SIGNIFICANT AMOUNTS. IT CAN BE GROWN IN DIRTY MEDIUMS SUCH AS POLLUTED WATER, RESIDUAL WATER, AND OCEAN WATER WITH HIGH CONCENTRATION OF SALT.

STUDY HAVE SHOWN THAT ALGAE CAN ABSORB 99% OF CO_2 IN WATER AND CONVERT IT INTO OXYGEN OR BIO - MASS. THIS TYPE OF FILTRATION PLAYS A MAJOR ROLE IN REDUCING CARBONDI- OXIDE IN ATMOSPHERE AND CLEANING INDUSTRY - POLLUTED AIR AND TO PROVIDE MORE OXYGEN IN WATER TO SUPPORT LIFE.

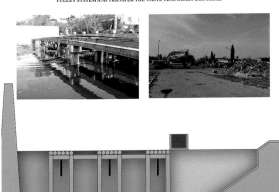
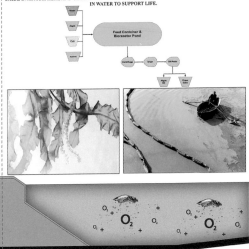

Passed through garbage from the existing filtration machine

The reservoir water is contaminated with waste of households and factories

Illegal settlement

The pluit reservoir

WHY?

Jakarta is still in the process of developing and is constantly changing. In its development, the city needs this project as a catalyst or example that helps in reaching its identity or what the city is planning to be. This project can **trigger the right changes** of Jakarta city development itself, before the progress stops and results in a bad product. Jakarta - as the metropolitan and capital city of Indonesia – puts currently rural development aside. **Lack of attention** to the stability **of rural areas,** as the production zone for a city's consumption, could eliminate the balance of city development. As the result of this **unbalance,** villagers begin to migrate into the city that results in the explosion of the city's population, the emergence of informal and **illegal settlements** and slums, followed by over-population, **lack of employment, poverty** and an increased crime rate, **poor waste management** and so forth.

REED FILTRATION SYSTEM	HYACINTH FILTRATION	WATER SPINACH FILTRATION	AQUAPONIC SYSTEM	REJUVENATION OF SPACE	SYMBIOSIS HOUSE
REED GROWS WILD AND OFTEN FOUND ALONG CALM AND MUDDY STREAM. ITS ROOTS CAN WITHSTAND MUD. THIS PLANT HAS A HOLLOW STEM, ROUGH AND HARD LEAF STALK. REED LEAVES ARE RATHER WIDE, SHARP AND POINTED AT THE END OF LEAVES TIPS. THIS PLANTS CAN GROW IN 6.6 - 26.6 CELCIUS AND 4.8 - 8.2 PH. REED HAS THE ABILITY TO DEGRADE ORGANIC MATTER CONTENT IN THE EFFLUENT. REED LEAVES CAN BE USED AS WOVEN CRAFTS MATERIAL AND TO CREATE TRADISIONAL MUSICAL INSTRUMENT.	WATER HYACINTH HAS 72.63% OF CELLULOSE THAT CAN BE USED TO ABSORB CERTAIN MATERIALS. THIS PLANT CAN PURIFY WASTE IN WATER WITH BIOFILTRATION TECHNIQUE AND ABSORB HEAVY METAL SUCH AS Pb, Cd, Hg, Zn, Fe, Mn, Cu, Ni, Au, Co, AND Sr. WATER HYACINTH IS ABLE TO ABSORB ORGANIC COMPOUNDS AND OTHER INGREDIENTS THAT MAY CONTRIBUTE TO THE EUTROPHICATION OF WATER BECAUSE OF THE ABILITY TO ABSORB NITROGEN AND PHOSPHORUS. WATER HYACINTH CAN DECREASE COD LEVEL UP TO 21.59% AND TSS UP TO 41.3%. OTHER USES OF WATER HYACINTH	WATER SPINACH PLAYS AN IMPORTANT ROLE IN KEEPING A BALANCE IN RIVER LIFE SUCH AS MOSS, PARAMAECIUM, AND FISH. THIS PLANT CAN NEUTRALIZE ALKALINE (K AND Na) IN POLLUTED WASTE WATER. K AND Na ARE THE MACRO ELEMENTS NEEDED BY WATER SPINACH TO GROW PROPERLY. WATER BECOMES CLEARER AFTER THE ABSORPTION AND WATER SPINACH ALSO BALANCING THE ACIDITY OF WASTE WATER AND KEEP IT ON 7.	AQUAPONICS IS A SUSTAINABLE FOOD PRODUCTION SYSTEM THAT COMBINES A TRADITIONAL AQUACULTURE (RAISING AQUATIC ANIMALS SUCH AS SNAILS, FISH, CRAYFISH OR PRAWNS IN TANKS) WITH HYDROPONICS (CULTIVATING PLANTS IN WATER) IN A SYMBIOTIC ENVIRONMENT.	AFTER THE CLEARANCE OF SLUM HOUSES ON ILLEGAL LOT, THIS LOT IS TURNED INTO GREEN SPACE THAT SERVES A GREATER PURPOSE TO SUPPORT ITS COMMUNITY. THIS LOT IS DESIGN AS: OPEN GREEN SPACE + RAIN HARVESTING SYSTEM + COMMUNITY AREA + PUBLIC TOILET + WASTE AND ENERGY TREATMENT THIS NEW DESIGN SPACE IS NOT ONLY GIVE ADVANTAGES FOR THE COMUNITY THAT WAS HERE BUT ALSO FOR EXISTING HOUSES ACCROSS THE STREET THAT NOW SERVE AS HOST FOR NEW HOUSING.	SYMBIOSIS HOUSE IS A NEW HOUSING SYSTEM THAT IS DESIGNED TO SIT ON EXISTING HOUSING ON LEGAL LAND. THIS HOUSING SERVES AS A NEW LIVING SPACE FOR COMMUNITY THAT STAY ON ILLEGAL LAND. THE FLEXIBILITY OF ITS DESIGN PREVENT ANY DESTRUCTION TO ITS HOST BUT IN THE OTHER HAND, SYMBIOSIS HOUSE PROVIDES WATER AND OPEN GREEN SPCACE FOR THE HOST.

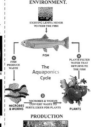

The **use of low tech and local tech** for the structures is the best way to counter massive building development in this environment. The Symbiosis House offers incitements to use and re-use recourses in a sensible way: bamboo as locally grown material for the building's design and rainwater, collected and stored for the host houses and the surrounding community. The concept of restoring the reservoir creates not only a healthy and comfortable place to live but symbolises a living example of **utilizing nature as technology** in balancing nature itself. The existence of an organism in nature has a distinctive role in its existence therefore with wise utilization of the environment **we can restore nature's equilibrium and natural ecosystems.** Thus, **learning from nature** is the most appropriate technology for reuse in terms of maintaining the balance of life.

WHAT?

New Jakarta Green Belt
An Urban Planning Vision, Jakarta, Indonesia

The New Jakarta Green Belt will provide low-mid dwelling for people and open public space, as well as green space which has been forgotten in Jakarta.

Office Atelier Cosmas Gozali, Jakarta Selatan, Indonesia
Location Jakarta-Bogor-Tangerang-Bekasi, Indonesia
Design period December 2012 - January 2013
Realisation period not yet
Concept by Cosmas Damianus Gozali
Team Rudy Hermanto, Michel, Jonatan
Website www.aryacipta.com
www.ateliercosmas.com

This project proposes a solution to the city government to make Jakarta a more decent and comfortably habitable "metropolitan city". The idea is to build a new ring belt outside Jakarta, the "New Jakarta Green Belt". The ring belt works like a multilayered bridge, 200 meter wide and is elevated above ground level to keep the land below as a water infiltration area and minimize environmental destruction. The belt includes infrastructure line for MRT, railways, vehicles, parking lots, mechanical electrical piping lines and water reservoirs. The top level is only for greening, bicycles and pedestrian pathways, which will increase the living quality and the open green space in Jakarta. It offers public housing on the platform with facilities like schools, hospitals, amusement parks, and shopping centres. This mixed use is for daily living needs, so the inhabitants do not need to travel far.

At the intersections of the belt with the existing main road around Jakarta, new interchanges are planned as satellite cities to decentralize city growth and to decrease pollution and traffic jam. The city centre will be more focused as a government, economic and cultural centre and thus will not increase the burden on nature and the environment. Citizen growth will be more controllable, security and comfort will be increased as well as the quality of living.

Although Jakarta is a waterfront city it is not perceived as such, like Singapore, Dubai, or Sydney. Tanjung Priok Sea Terminal will be moved outside the ring belt to improve the waterfront. This waterfront city will also be connected to "Seribu Island" which is located 45 minutes drive north of Jakarta and will become a cultural centre, maritime museum and resort area. New Jakarta airport will be moved to the northern seashore side of Tangerang city to avoid a legal dispute for land acquisition when the government plans to expand the airport. All these "New Waterfront Cities" can be sold at a very good price in order to finance the overall project.

With the New Jakarta Green Belt interchange stations will be generated which are the starting point of decentralization in Jakarta. They decrease the movement of people from one place to another. This also stimulates the development of a new economic centers in the surrounding area.

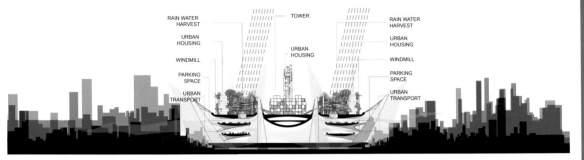

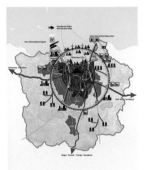

The "New Jakarta Green Belt" decentralises Jakarta – a solution for issues that occur nowadays such as flooding, traffic jams, pollution and slum areas. The city will spread into several areas around Jakarta and there will be an interchange point connecting from one area into another, so it becomes a circle line.

Public housing along the belt will provide a new home for illegal citizens who occupied most of the riverbanks and turned them into slum areas. Riverbanks will regain their initial function as green and water infiltration areas, river pollution stops and green areas in Jakarta will increase from nowadays 7% to 40%.

A rainwater catchment dam will conserve rainwater. The excess will be distributed through piping line in the new ring belt and kept in water catchment areas in the green belt for the use of the public housing.

HOW?

A New Green Metropolis on the northern coast of Jakarta with a sustainable environmental concept. The New Jakarta Green Belt will increase the economy for Jabodetabek-punjur with recreational facilities, a cultural center, resort and other supporting facilities. The dam at north Jakarta, which also serves as circulation, could control sea levels so the flood can become a power generator.

WHY?

The uncontrolled growth of slums in some areas like along the riverside and along the train railway. These cause the decrease of the green areas in Jakarta.

Traffic jams always occur at almost every main road. Especially in the morning and afternoon. Traffic jams can't be avoided even on the freeway.

Jakarta is one of the rapidly growing cities in the world, both in terms of economy and population, and a magnet for most people. As a result the **population has become uncontrollable,** the city has **many illegal citizens** and added problems. Starting from the settlement problem, **other problems arise, like infrastructure, waste treatment, green environment, social economic and security.** Environmental destruction and bad waste management have become the main cause to catastrophe and flood.

Jakarta has so many issues such as: • High traffic density • Flooding • Lack of green areas • The growth of slum areas in many points • Air and water pollutions • Bad mass transportation system • Rivers not functioning properly

Jakarta needs the "New Jakarta Green Belt" system to solve its complex problems. Indonesia needs to catch up with other countries without losing its local identity but preserving its culture without influences of other countries. Indonesia will emerge as a developed country with its own character.

Areas are always flooded after heavy rain. Floods always cause traffic jams, distracting from economic and social activities for Jakarta people and causing sickness afterwards.

The weak control of industrial waste and waste management in some areas of Jakarta.

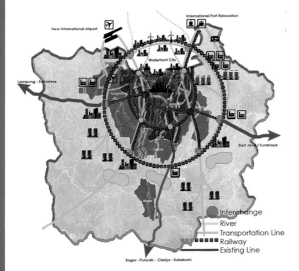

The New Jakarta Green Belt will restrict the number of vehicles coming in and out of Jakarta with interchange stations at several points. Public transport will be spread around the city to reduce the use of private transport and pollution.

The river will be used as an alternative transport route, especially in the main city of Jakarta, in order to reduce the amount of land transportation. Along the riverside water recreation will be supported.

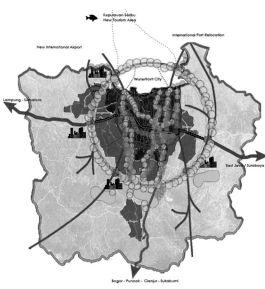

To restore the function of green open areas along the riverside as the water infiltration system. This green open area will function as the recreation and excursion area, promenade, biking zone, and other functions. And it will be supported by the use of the river as water transportation.

Mangroves will be preserved along the northen coast to reduce abrasion, to increase green open area and conservation forest, and can be used as a forest research centre and recreation area.

Build a dam at the Northern coast of Jakarta, which is not only a connecting road, but also used to control the sea water level to prevent flooding and to gain new hydro-power resources.

Build the water reservoir to prevent the city from being flooded. The water reservoir system is connected by pipes along the track of New Jakarta Green Belt which flow into the sea.

The citizens will learn to live sustainable by **using alternative energy resources (wind and solar), public transportation to preserve the environment** and will be **encouraged to socialize and live vertically** as the living space will be limited and vertically raised.

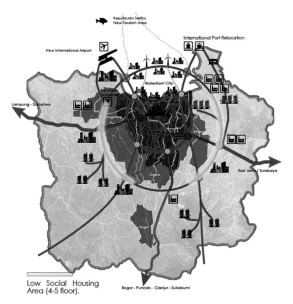

Low Social Housing Area (4-5 floor).

The border between the New Jakarta Green Belt and the northern coast of Jakarta will be developed as the new commercial & real estate, entertainment, resort, cultural center and other functions.

The new International airport will be built at the northern coast of Jakarta with coast reclamation to avoid land property disputes, while the old airport will be used for domestic flights.

Tanjung Priuk port will be relocated north, outside of New Jakarta Green Belt, and will be developed as an international port that will compete with other international ports around the world.

The illegal housing along the riverside will be relocated to the low-mid dwellings at the New Jakarta Green Belt.

WHAT?

Alternative Helicopter: Ayun Ayun Kaliku

Enhancing the Awareness of Endangered Polluted River, Kampung Cikini Kramat, Central Jakarta, Indonesia

The Urban Intervention Workshop is a collaboration of Indonesian and Japanese Universities, which has been running since 2011. The workshop is a creative experiment and looks at alternative methods of urban intervention by bridging the gap between global communities and local initiatives in particular. In the first two years the workshop focused on Kampung Cikini-Ampiun. This is a four hectares district next to the central business area of Jakarta with a population of about 3,200 people. It is one of the strategic areas of Jakarta, which implies also uncontrollable growth without any innovative approach.

Through the highly populated neighbourhood of Kampung Cikini-Kramat Ampiun runs a small river, a side stream of the Ciliwung. The stream is practically an open sewer and residents are worried about the water quality and contaminated water in the wells nearby the river. One reason for the river's pollution are self-made public toilets, the so-called "WC Helicopter", located directly on the more than ten bridges crossing the stream and the toilets' waste goes directly into the stream. Most of them have been abolished and replaced by proper public facilities including a well, baths and toilets, called Mandi Cuci Kakus (MCK).

Architecture students from Indonesia and Japan developed design ideas for urban interventions in 2011 and 2012, and shared them with the local community. In the second Urban Intervention Workshop in 2012 they proposed to the residents their explored alternatives for the two remaining 'WC Helicopters'. The workshop's outcome finally was the temporary installation of a bamboo swing: Alternative Helicopter Ayun Ayun Kaliku.

Organizers Research Institute for Humanity and Nature (RIHN), Japan Universitas Indonesia, Chiba University, Jakarta, Indonesia
Location Kampung Cikini Kramat, Central Jakarta, Indonesia
Design Period 6 - 16 September 2012
Construction/Realisation period September - October, 2012
Concept by Evawani Ellisa and all project teams
Team Participants: 8 students (Universitas Indonesia), 5 students (Chiba University), 1 student (Kyoto University) Coordinators: Akiko Okabe (Chiba University), Evawani Ellisa (Universitas Indonesia), Tomohiko Amemiya (architect, Tokyo Metropolitan University)
Website www.weuhrp.iis.u-tokyo.ac.jp/ chikyuken/publication/webdir/2/ SensibleHighDenCity.pdf
www.youtube.com/watch?v=Z622sbGAaZA

Opening ceremony

Discussion with community

The installation of the swing is a **participatory project** as it has been selected, developed and built together with the local community. The swing is a statement and integrated response to the community's threats and wishes in terms of water **quality and garbage collection.** Moreover the **swing functions as a mini-playground** as the community lacks public spaces and playgrounds, due to intense urbanization and motorbike parking.

Preliminary ideas

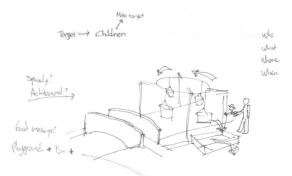

Construction process

HOW?

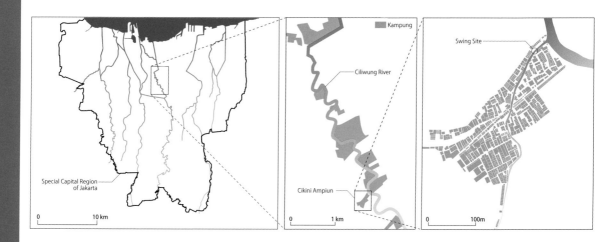

WHY?

Jakarta, a megacity with more than 10 million inhabitants, has a sewage system, which covers an actual area of under 3%. Wastewater in the densely populated areas of central Jakarta has always been drained into the natural river system and the district Kampung Cikini-Ampiun has one of those **polluted rivers** passing through. The swing was installed with the objective to **enhance awareness** of the endangered polluted river; moreover it symbolizes a participative urban intervention in a district threatened by redevelopment and rapid growth.

If Kampung Cikini-Kramat Ampiun were to be redeveloped, **the actual residents with different income statuses would be pushed out** from the city centre and not able to maintain their quality of life. The swing as a low-tech urban intervention encourages the community to stand exemplarily for sustainable urban development, which **improves living conditions,** respects the urban fabric and prevents social divide.

The condition of the existing river Kali Keroncong

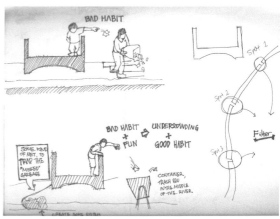

Idea to change the habit of throwing garbage into the river

The installation is expected to **function as a trigger for the community** to turn the river into a place for relaxation and a playground for children as a long-term vision, far beyond the existence of the swing installed by the students' project. That transformation would **require the residents' participation, solidarity and a certain public spirit.** As a next step the community could start the initiative of not only improving their living environment but also **reclaiming the water ecosystem** in their neighbourhood.

Garbage collected during event

At the opening ceremony more than 100 people, children and local leaders, gathered together. Children formed a long line half an hour before the opening ceremony, following the motto: "You can take a first ride on the swing if you **come with garbage instead of throwing it into the river!**" The laughter of children playing with the swing persisted till mid-night. The swing itself only lasted for three months, as it was not designed to last for a long period. But that was quite pleasing as it exceeded the estimation that it would only last for one month.

WHAT?

Children standing in line to exchange their garbage for a ride on the swing

Workshop participants and children of kampung community

Opening ceremony

Fluidscape City

Urban Development for Contemporary Reservoir City Project, Penjaringan, North Jakarta, Indonesia

Exterior view of flying hotel and multipurpose hall

Office Budi Pradono Architects, Jakarta, Indonesia in collaboration with Bina Nusantara University & Tarumanegara University
Location Pluit, Penjaringan North Jakarta, Indonesia
Design period December 2012 - April 2013
Construction/Realisation period unbuilt
Concept by Budi Pradono
Team Budi Pradono Architects Team: Stephanie Monieca, Hasan Nuri, Anggita Yudhisty Nasution, Reini Mailisa, Intan Kusuma Dewi, David Kurnia, Awly Muhammad Isra, Riangga Yudastira, Lucia Wili Yuhartanti
Tarumanegara Team: Priscila Epifania
Student: Guntur Haryadi Halim, Raisa Hakim, Indra, Martin Alvin Setia Ekacahya, Gratio Ray Sutanto.
Bina Nusantara University Team: Firza Utama
Student: Davin, Rheza Maulana, Bakrie
Film Production: Genesis Production www.facebook.com/gensfx
produced by Bagus S. Pradono
Putri Utami - creative
Andria A. Putra - graphic designer
Warsito - crew
Anton jr - editor, Animator
Kevin MacLeod - music creator
Suwarto Kartasoewarto - video footage
Photographer: Jonathan Raditya, Reynaldo Tjandra, B.Jesica Valeria, Martin Alvin Setia Ekacahya
Website www.budipradono.com

Fluidscape City is a study of visioning the future of Pluit Reservoir. The development of the area consists of 280 ha of waterfront area, which includes 80 ha of reservoir from a major river crossing Jakarta. It was built to anticipate the flood by the Dutch in the 18th century.

The reservoir functioned well in defending against the flood, until this area was used as a garbage dump and people began to build illegal houses on it. Increasingly illegal houses have covered 20 ha. The reservoir has been neglected. They started to protrude towards the reservoir, with waste disposal into the river. Siltation and the disposal of industrial wastes made it dirty, smelly, and unhealthy. Problems were emerging; there has been major flooding every 5 years, the latest occurred in 2013. At that time there was damage to the pump reservoir which resulted in even worse flooding.

Some questions to be considered are: Could this be a new device for reservoir development in Pluit area? With the latest technology do we need to create a contemporary sustainable windmill, which is able to regulate water flow in anticipation of the flood? Could this be part of the area of new business development that relies on water as a source of energy, water resources, as well as a new tourist destination? What about the slums, how do we sustain the illegal population at the same time as supporting urban life in the region? How can new development move all economic activities whilst at the same time enabling people to prosper? What if we create an urban vertical farm integrated with water filtration windmill-pumps with more sophisticated devices?

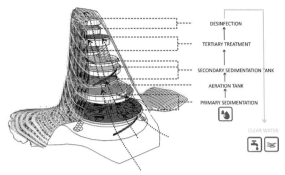
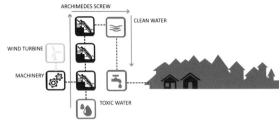

Water treatment diagram The turbine speed is adjustable, so when floods happen the turbine gets faster and so does water distribution. This system is expected to prevent the reservoir from flooding.

Futuristic ecological windmill with archimedes screw and fan (water treatment diagram)

View of water treatment tower with archimedes screw

Interior of visionary flying hotel

This project is an attempt to respond to catastrophic floods in Jakarta and to preempt further flooding with **a concept that turns the disaster into a new force** to produce a contemporary 'smart city'. The area around the reservoir is designed as a new central business district, equipped with a future pump building. Using the Archimedes method that relies on a lot of fans controlled by the wind and the sun, this pump station, which regulates water management and makes biological filtration, can be used to **treat water for the citizens of Jakarta and irrigate the whole region. A vertical urban farm will supply organic food across the region.** The vertical kampung (social housing) will replace the slums whilst retaining their spirit and create balanced communities in the surrounding areas. It will utilize water as the driving force of its energy and at the same time take advantage of the building skin to absorb heat throughout the year. The reservoir, besides functioning again as a means to distribute water in Jakarta, can also function as a future tourist area. The alternative building typologies and a new-found specificity of place in the urban context of Jakarta will deliver added value for the Fluidscape City character.

HOW?

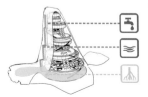

Water use and Mangrove diagram

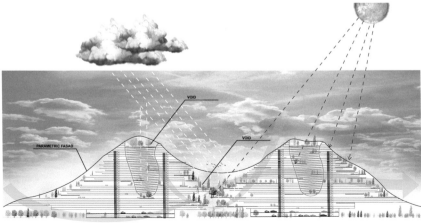

Future apartment building with ecological approach

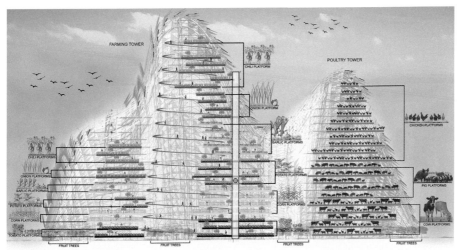
Urban farming schematic diagram

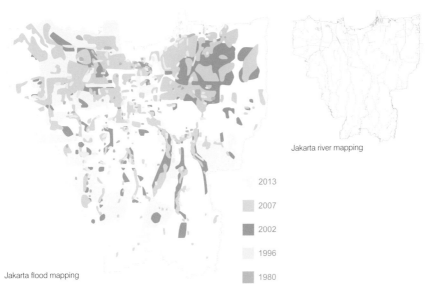

Jakarta river mapping

Existing landuse diagram

2013
2007
2002
1996
1980

Jakarta flood mapping

Potential development diagram

WHY?

From year to year flooding from the Ciliwung river flow is getting bigger, due to the swamp loss and fast-paced, unsustainable urban development. **Water has begun to destroy the city because of the imbalance of natural forest management and rapid development, and the number of people, cars and concrete buildings.** This project anticipates future natural disasters and utilizes new technology to manage the floods as a force for generating energy.

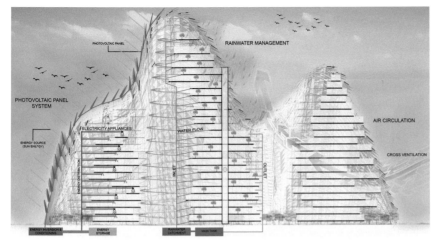
Diagram of Climate management system with photovoltaic responsive skin, energy storage, and rain water management.

This will provide an opportunity for the community to enjoy the beauty of the waterfront area and **create awareness of the importance of water to human life.** People are forbidden from throwing any waste into the river. Education needs to be established to respect the water. **The vertical urban farm gives local people the opportunity to adopt the latest technology** in the use of hydroponic plants. People can also do tourist activities, sport and exercise, as well as **gain new knowledge about nature.** The Fluidscape City concept will smarten the city and encourage people to believe in the new technology and perceive the spirit of the place in a different way.

WHAT?

Future office tower at Fluidscape City

Future waterfront Fluidscape City

Future waterfront Fluidscape City

Pajak USU
Market Revitalization at the North Sumatera University, Sumatra, Indonesia

Pajak USU is a market at the campus of the University of North Sumatra (USU) in Medan, the third largest city in Indonesia, located on the island of Sumatra. A group of independently working Indonesian architects formed a network to develop a new concept of traditional market. The result is a two-story market building with gallery space and semi-permanent shops, crafted in collaboration with the existing community of merchants. With the focus on sustainability the project includes local materials, rainwater recycling and alternative energy production.

Originally Pajak USU was just a bunch of street vendors along the road selling mainly to university students. When the monetary crisis hit Indonesia in 1998 a rising number of hawkers started a flourishing business and the market became more and more attractive for other traders and city dwellers. The market conditions since then became increasingly irregular in terms of hygiene, water and energy supply, the crime rate rose, traffic went out of control and the image of the university suffered. Once the stalls had to be rebuilt by the merchants after a fire in 2009, USU started to consider removing the market from the campus based on the university's land-use plan, which does not allow markets within the site.

The project is an attempt to mediate between the community of merchants and the university and to create a partnership in order to revitalise the market site.

Office Tim Pajus, Medan, Indonesia
Location Medan, Sumatra, Indonesia
Design period 2010
Construction/Realization period unbuilt
Concept by Ramadhoni Dwi Payana
Team Rangga Mury, Rudi Hermanto, Rahardian Pradityo
Award First Price Futurarc International Competition 2011

The merchants, thanks to their crafts and their knowledge of local and affordable materials contribute to the building's design concept and selection of material. In contrast to the existing semi-permanent buildings, constructed after a fire by the merchants themselves, Pajak's new development is supported by the university, as the market's services are highly demanded by USU students and the communities around.

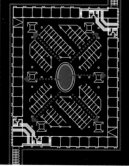

The layout of the semi-permanent shops provides good visibility of the site and allows visitors easy access. The central plaza serves as a meeting and viewpoint.

The traditional market is the heart of the micro economy in Indonesia. All parts of society come and interact there. The project brings together design, technology, participation and sustainability:

Initiative and partnership
The project takes the merchants' initiative into consideration and attempts to **restore the spirit of their community by funding and rebuilding the market in partnership** with them.

The use of local materials
Locally produced construction materials facilitate the project's implementation and the market's maintenance. Bamboo and palm trees are grown in-and outside the market **as sustainable building materials.**

Recycling
Rainwater is recycled through a filter system in the market's roof structure and is re-used by the market community. **Biogas is produced** as renewable energy out of the market's organic waste and transformed into fuel f. e. for cooking. The market's roof structure provides natural light and ventilation to **reduce electricity consumption.**

Limitation of the market's site
Each floor and the outside area are defined for certain goods and services to avoid that the market grows irregularly.

Circulation
By providing two main entrances and separate ramps for pedestrian and motorcycles, **problems like congestion and cross circulation are minimized.** Food stalls are located on the 2nd floor for hygiene reasons (insects).

Fire Safety
Rainwater is collected in underground basins and can be used as a first step to fight fire.

HOW?

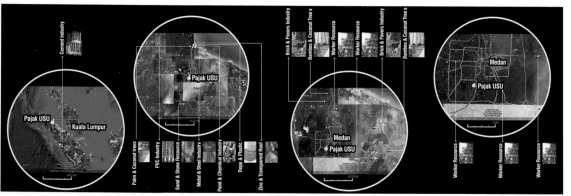

Material resources

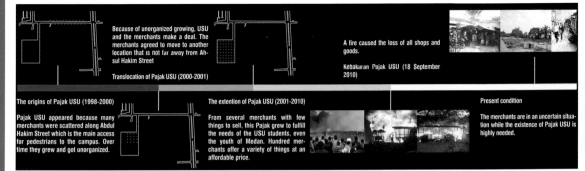

The origins of Pajak USU (1998-2000)
Pajak USU appeared because many merchants were scattered along Abdul Hakim Street which is the main access for pedestrians to the campus. Over time they grew and got unorganized.

Translocation of Pajak USU (2000-2001)
Because of unorganized growing, USU and the merchants make a deal. The merchants agreed to move to another location that is not far away from Ahsul Hakim Street

The extention of Pajak USU (2001-2010)
From several merchants with few things to sell, this Pajak grew to fulfill the needs of the USU students, even the youth of Medan. Hundred merchants offer a variety of things at an affordable price.

Kebakaran Pajak USU (18 September 2010)
A fire caused the loss of all shops and goods.

Present condition
The merchants are in an uncertain situation while the existence of Pajak USU is highly needed.

Time line

WHY?

Traditional markets in Indonesia are identified with clutter, traffic jams, shopping discomfort and crime. However, those markets still remain attractive because they are affordable and places of social interaction. Markets built by the government often lack understanding of social and climatic conditions and ignore energy saving aspects, but focus on economical aspects only. **A market as a social space and responding to questions of sustainability is needed.**

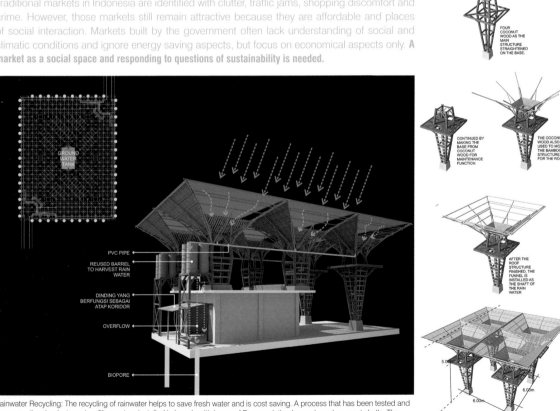

Rainwater Recycling: The recycling of rainwater helps to save fresh water and is cost saving. A process that has been tested and proven as the simplest way is a filter system installed in barrels with layers of Dacron cloth, charcoals and coconut shells. The recycled water is used in the market eg. for washing, watering plants and in toilets and contributes to a sustainable way of living.

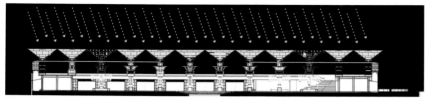
Rainwater Harvest

The entry area is dedicated to the USU students. The space is designed as a gallery to host events like performances and exhibitions to showcase, promote and sell students' work and improve students' entrepreneurship.

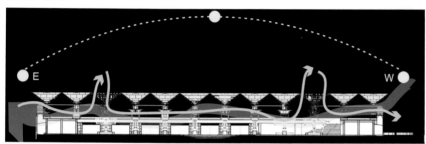
Wind and Sun

The inner court functions as a central point of orientation of the site and the building circulation, and also as a public space for the people.

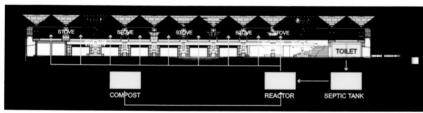

Biogas is a renewable energy source and is produced from regionally available raw materials and recycled waste, such as manure, sewage, municipal waste, green waste, plant material, and crops. It can be used as fuel for example for cooking for the food vendors.

Open areas in- and outside the market are transformed into public bamboo and palm tree gardens, which are maintained and used by the merchant community to grow sustainable building materials.

The design seeks to provide equal opportunities for all stall holders to get a place that is easy to reach and seen by all people. Visitors can see in all directions when entering the market. Small stalls that occur in the middle part are, due to security regulations and to enforce visibility, not more than 1.5 m high.

The inner court in the market serves as a centre for student activities. Students learn for example entrepreneurship through a free stall given to each department; they exhibit and organize art events to the benefit of all customers and the market community.

Side elevation

WHAT?

Smart City Workshop - Jakarta
January 2013, at Bank Mandiri Museum, Jakarta, Indonesia

Young professionals from the fields of architecture, design, sociology, and other disciplines, under the guidance of Rujak Center for Urban Studies Jakarta, and the Goethe Institut Jakarta met at the Bank Mandiri Museum to imagine and debate the future of Jakarta. The museum is housed in a heritage building in Jakarta Kota (Jakarta Old Town) previously the first headquarters of the Netherlands Trading Society. It was in this special context in between a collection of colonial banking-related items, the historic environment of the former core of the city, close to the large harbour kampung and the most crowded traffic situation around Jakarta Kota Station, where twelve participants and numerous invited guests found the perfect location to rethink the city of Jakarta. A framework of important issues in regards to urban behaviour structured the workshop theme. How can new smart strategies change the behaviour of citizens? To understand how to influence the forces, which lead to the actual physical appearance of Jakarta the workshop team visited several sites and interviewed citizens in their local context including Kampung Pulo and Kampung Bukit Duri in East Jakarta. Furthermore expert lectures and briefings prepared the ground for a set of imaginative projects.

Jakarta, the capital city of Indonesia, with a population of 28 million people in greater Jakarta represents the most populous city in Southeast Asia. The growing population of Jakarta, its unlimited growth, and the large amount of commuters between travelling everyday between working and living together with the consequences of its fast economic growth of 7% over the last year 2012 became subjects of investigation for ten days. A catalogue of nine urban disasters happening in regularly in Jakarta became an ongoing major concern among the team members during the workshop. Solving these issues comprising floods, fires, earthquakes, tsunamis, epidemics, social unrests, extreme climates, extreme tidal waves, and technology failures in relation to urban behaviour became a challenging task and focus point. "Prepare the umbrella before it rains", a famous Asian saying, represented an overall Leitmotiv for a new lifestyle of awareness and preparation. The project "14% Green Space" proposes to enlarge the amount of green areas to reduce annual floods as a direct answer related to this discussion.

Guided by Elisa Sutanudjaja, Rujak Center for Urban Studies, Jakarta and Dietmar Leyk, Leyk Wollenberg Architects, Berlin; Visiting Professor at The Berlage Center for Advanced Studies in Architecture and Urban Design - TU Delft

Students participating in the workshop were Andreas Yanuar Wibisono, Dzikri Prakasa Putra, Syarfina Mahya Nadila, Elbert Chayadi, Mochamad Hasrul Indrabakti, Mufty Riyan Firmansyah, Muhammad Fatchurofi, Ria Pratama Istiana, Anita Halim Lim, Robin Hartanto Honggare, Soitua Sidjabat, Fariduddin Atthar

January 3rd 2013
Introduction: Smart City and Frightful Facts of Jakarta
Presentation of proposals and feedback from Dietmar Leyk and Rujak's team: Marco Kusumawijaya, Dian Tri Irawaty and Elisa Sutanudjaja.

January 4th 2013
Mobility, Habitat, Water-Related Issues and Disaster Lectures by Suryono Herlambang, Yoga Adiwinarto, Elisa Sutanudjaja, Selamet Daroyini and Hening Parlan
Discussion with public and participants

January 5th 2013
Walking in the City Session about street photography by Erik Prasetya, one of

the best street photographers in Asia. Tour of Urban Poor Network's kampong in Muara Baru, North Jakarta and to an urban poor settlement along Ciliwung River in Bukit Duri, South Jakarta. Meeting with the community leaders, Sandyawan Sumardi and Ivana Lie from Ciliwung Merdeka.

January 7th 2013
Middle-Class Community and How They Thrive in Jakarta Presentation of three different middle-class communities of ideas and strategies in solving problems.@nebengers, carpooling; Green Community of Pondok Indah; Shanty Syahril, after-school program for neighborhood kids and library for urban poor children.
Discussion of the projects and work on proposals

January 8th - 9th 2013
What is Normal? Final lecture with Khairani Barokka
Work on proposals and presentation of ideas

January 10th 2013
Critique from Practitioners Professional input by: Hizrah Muchtar, urban planner; Avianti Armand, architect and Bayu Wardhana, social entrepreneur.

January 11th 2013
Final Presentation with Famega Syavira, journalist Yahoo! Indonesia

January 13th 2013 Public Expose @ GoetheHaus Auditorium
Public discussion with representatives from the Ministry of Public Works, the Ministry of Transportation, State Ministry for the Environment, the National Planning Board, and Jakarta administration.

The emphasis of the smart city exploration was on intelligent solutions for environmental, infrastructural, social and sustainable issues in urban contexts. The hypothesis stated, that the future Jakarta is smart, meaning: it is about intelligent, integrated, and networked strategies.

A constellation of 12 smart projects acts as first frames to define the city in a new way. The projects combine the two extreme scales of the city, the metropolitan and regional geography (Magic Belt) and the architectural intervention (Berbagi Ruang: A Space Sharing Platform, which awakes a collective awareness to share space in a growing metropolis and to redistribute what already exists).

The greater amount of the projects deals with issues in regards to Jakarta public transport. Trans Jakarta Network and Jakarta MRT are not only subjects of daily political debates and media, but belong to the very usual every day experience of all citizens. Mobility in Jakarta comprises forces, which lead to almost unbearable commuting conditions and dominate the daily life performance of the Jakarta citizen. These projects contain very strategic aspects to maximize connectivity, to travel faster by a safe, comfortable transportation system and at the same time to reduce vehicle trips and the miles travelled by the large amount of private motorbikes and cars. "Carnival Mobility" shows a fascinating example how a train ride becomes more enjoyable though interactive behaviours.

Since Indonesians are among the world leaders in using all kind of advanced digital communication devices, using new applications for relevant social projects seems obvious. Out of this comes the idea to create the "Steady Family" mobile application, a crowd-source-based app that assists families to be pro-actively prepared against the flood, trigger home scale solutions, share smart tips in dealing with floods and environments as they spent quality time with their family. Another proposal suggests how to use media to connect existing communities in Jakarta and beyond. A very interesting debate developed around the question how to control all these proposals. "Jakarta Besok: Start from a Dream" suggest an interactive website where citizens can monitor the progress of transforming Jakarta. They can agree (amen) or disagree (boo) with the idea or the evolution of a project. "This project is not about fiction, dreams, or wishful thinking, but about future ideas for Jakarta. It serves as reminder for people and the government".

The smart city workshop Jakarta itself and due to its public presentations and debates, among them the impressing final presentation at the Goethe Institut in Jakarta, succeeded in creating the departure point for a common public awareness for new ways of smart performances and behaviours by raising challenging questions and creating very unexpected solutions for a desirable future for a Southeast-Asian city like Jakarta.

Dietmar Leyk April 2013, Berlin

Low-Cost/Low-Tech: Budget Air Travel and the Future of Southeast Asian Cities
A Research on Mobility

Low-cost carriers such as AirAsia, Lion Air, Nok, and Vietjet have radically expanded the socioeconomic spectrum of Southeast Asia's flying public. With slogans like "Now Everyone Can Fly," they offer cut-rate fares between the region's cities.

Office Max Hirsh and Anna Gasco, ETH Zurich / Future Cities Laboratory in Singapore
Locations Bangkok, Batam, Cebu, Ho Chi Minh City, Johor Bahru, Kuala Lumpur, Phnom Penh, Singapore
Research period 2012/2013
Website www.christiaanse.arch.ethz.ch

Across Southeast Asia, air traffic is growing at a prodigious rate. The rise of low-cost airlines has enabled millions of budget tourists, migrant workers, retirees and students to join the international flying public. At the same time, the development of air cargo networks in Southeast Asia's hinterlands has allowed small and medium enterprises to extend their commercial activities across national frontiers.

With that in mind, Low-Cost/Low-Tech investigates how Southeast Asian airports are negotiating vastly enlarged flows of goods and people. It seeks to convey the inimitable mixture of excitement, chaos and discipline that pervades airports in the region. At the same time, however, the boom in air travel has been accompanied by a heightened awareness of aviation's deleterious consequences. Southeast Asian airport operators are confronted by two contradictory impulses: a celebration of the democratization of air travel and greater freedom of movement; and the desire for more rigorous environmental protection and sustainable urban design.

Low-Cost/Low-Tech documents how a variety of actors--airport authorities, architects, logistics firms, entrepreneurs and passengers--have negotiated the exponential increase in air traffic in places where, not so long ago, aviation was the reserve of a select few. In so doing, it uses airport infrastructure as a lens for interrogating a much larger question: How are Southeast Asian cities being redesigned to accommodate vastly expanded flows of goods and people, who are on the move for the short-, medium-, and long-term, and who operate on hugely diverse socioeconomic levels? Ultimately, it argues that effective engagements with Southeast Asia's future mobility challenges will rely on a creative combination of both "high-tech" and "low-tech" solutions and that it is crucial to study their interaction in order to anticipate new opportunities in the region.

The infrastructure of low-cost air travel in Bangkok, Ho Chi Minh City, Kuala Lumpur, and Singapore.

Over the past decade, many cities in Southeast Asia have built enormous new airports, located far from the city, that were intended to replace smaller airfields within the urban core. Recently, however, many of these older airports--such as Bangkok's Don Mueang and Kuala Lumpur's Sultan Abdul Aziz Shah--have been reinvented as hubs for low-cost passenger and cargo flights. In the meantime, other airports such as Singapore's Changi have responded to the boom in low-cost travel by building new, no-frills terminals.

Budget air travel has radically **democratized cross-border mobility** in Southeast Asia. For many passengers, these flights represent the chance, unavailable to their parents and grandparents, to experience life beyond the confines of their home country. Yet while affordable, mass air travel opens up **new opportunities for work and leisure abroad,** as well as a broader understanding of the world, it also entails undeniable environmental consequences in terms of noise and air pollution; and places new stresses on urban infrastructure systems. This project investigates simple yet efficient forms of mass transit that have developed in Southeast Asian cities to accommodate flows of international passengers that are both much more numerous as well as much more **socioeconomically diverse.**

Migrant workers throughout Southeast Asia rely on low-cost carriers in order to travel back and forth between their hometowns and places of employment. The presence of these airlines is particularly noticeable in small towns and provincial cities that are not typically associated with the geography of international aviation.

WHY?

Many passengers in Southeast Asia **lack the basic infrastructure** that is needed to fly – such as a credit card, internet access, a way to get to the airport, even a last name. These pop-up infrastructure systems have in response developed a system for accommodating this new clientele of nouveaux globalisés. Moreover, a simple expansion of existing city-to-airport transport networks – which are largely car dependent – is not an option in the context of air travel's mass expansion. An investigation into sustainable, and affordable forms of mass transit is thus critical to the future development of cities across Southeast Asia.

The movement of cross-border cargo between Batam and Singapore.

Batam is one of the most important sites of production and transshipment in Indonesia. At Batam's Hang Nadim International Airport, a wide range of goods--from lettuce to microchips--is flown via turboprop planes to Seletar, a small airfield in northern Singapore. From Seletar, the goods are transported by truck to Changi, where they are distributed to consumers around the world.

In Southeast Asia, contemporary urban development relies on an intense combination of both "high-tech" and "low-tech" design solutions. It's not a simple case of one replacing the other – and it's crucial to understand their interaction in order to anticipate new opportunities for architects working in the region. An investigation of relatively simple, yet ultimately quite effective, modes of geographic displacement that are developing through low-cost air travel thus serves as a useful lens on the much broader topic of Southeast Asia's future urban development.

Many budget airlines have opened "pop-up" check-in centers, located in downtown shopping centers, where passengers who don't have access to the internet can buy plane tickets and find affordable city-to-airport transportation. In countries like Cambodia and Laos, the most economical means of getting to the airport is usually via tuk-tuk.

WHAT?

Fiber Composite Reinforced Concrete
A Material Research, Tropical Zone Worldwide

Close-up view of Bamboo Composite Reinforcement Material

Office Dirk E. Hebel, ETH Zurich / Future Cities Laboratory in Singapore
Location Tropical Zone World-Wide
Design period 2012/2013
Concept by Dirk E. Hebel
Team Felix Heisel, Alireza Javadian, Mateusz Wielopolski, Karsten Schlesier, Marta Wisniewska, Tobias Wullschleger
Website www.hebel.arch.ethz.ch

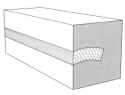

1. natural bamboo in fresh concrete

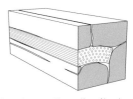

2. cracks caused by swelling of bamboo

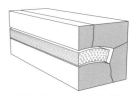

3. dried bamboo in expanded voids debonded from concrete mix

The effects of swelling and shrinking of natural bamboo in concrete as reinforcement

Newly developed organic composite materials as reinforcement systems in structural concrete have the potential to revolutionize the concrete building sector, which, over the last 100 years, has not changed significantly. Steel reinforced concrete is the most common construction material worldwide. However, very few countries have the ability and resources for the production of steel reinforcement. Even highly developed nations such as Singapore have to import all construction steel used for their rapid development. Secondly, the production of steel at extreme temperatures requires vast amounts of energy and resources. Shipping the material around the world afterwards only adds to the immense carbon footprint of construction steel.

The benefits of using bamboo as an alternative reinforcement material for structural concrete applications are numerous: fast growth, high tensile strength, and the capacity to capture large amounts of carbon dioxide from the atmosphere are just some of the most remarkable properties this grass has to offer. Further, the natural habitat of bamboo overlaps with the location of the fastest growing populations and cities worldwide and, thus, with the demand for reinforced concrete as a building material. This current research investigates the possibility of using fiber composite reinforcement as a renewable substitute for steel. If successful, this alternative technology could liberate construction sectors in developing territories from its dependency on heavy steel imports. In most developing territories, bamboo (as a source of high-tensile organic fibers) is a widely spread and easy to obtain natural resource. Local markets and value chains could be established by introducing an alternative and green technology which would strengthen local production and therefore create new job opportunities and by establishing social equities inside the countries.

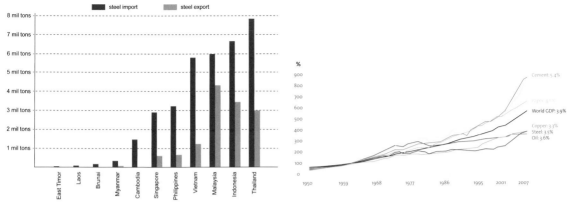

South East Asia Import and Export of Construction Steel 2010

Material consumption world-wide

By replacing steel with an organic and local alternative such as bamboo, **Fiber Composite Reinforced Concrete has the potential to smarten-up our cities in rapidly developing territories economically, environmentally and socially:** As a local resource, fiber reinforced concrete **reduces dependencies on foreign market prices** and import taxes. The necessary production steps **provide future cities with work and income** and **sustain farmers** and the environment equally. Unlike steel, bamboo is an organic resource, a rapidly growing grass, which **binds CO_2 in the production process.** Furthermore, while currently building typologies are being imported parallel with building materials, **a local and independent industry might diversify architectural styles,** following the social, cultural and traditional needs of each specific location.

HOW?

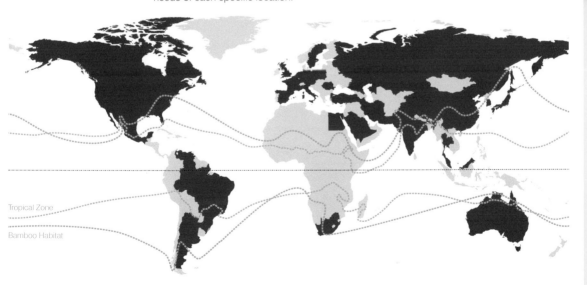

Tropical Zone

Bamboo Habitat

Countries with an annual steel production rate higher than 4 m tonnes Index Mundi, 2012

WHY?

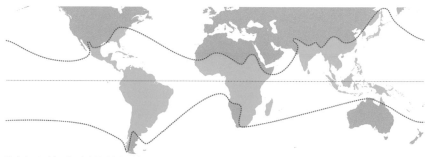

Global natural bamboo habitiat National Geographic, 1980

Bamboo's natural habitat can roughly be located along the equator. Most developing territories today with accelerating rates of population growth and increasing housing requirements are found in exactly the same geographical location, including South-East Asia. By introducing an alternative green technology to the urban production process, bamboo could strengthen income possibilities also in rural areas by forming local value chains. As a result, a reduction of migration tendencies towards cities might occur. It also offers a chance to combine and therefore reevaluate globally applied building materials with locally available resources and knowledge from the South. It is proposing a 'reverse' or 'alternative modernism', whereby developed countries and cities might learn and gain from the knowledge developed in the 'South'.

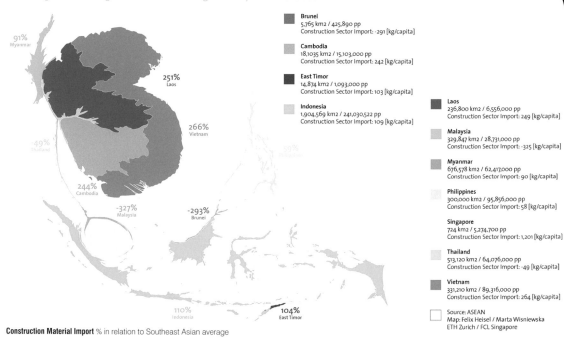

Brunei
5,765 km2 / 425,890 pp
Construction Sector Import: -291 [kg/capita]

Cambodia
18,1035 km2 / 15,103,000 pp
Construction Sector Import: 242 [kg/capita]

East Timor
14,874 km2 / 1,093,000 pp
Construction Sector Import: 103 [kg/capita]

Indonesia
1,904,569 km2 / 241,030,522 pp
Construction Sector Import: 109 [kg/capita]

Laos
236,800 km2 / 6,556,000 pp
Construction Sector Import: 249 [kg/capita]

Malaysia
329,847 km2 / 28,731,000 pp
Construction Sector Import: -325 [kg/capita]

Myanmar
676,578 km2 / 62,417,000 pp
Construction Sector Import: 90 [kg/capita]

Philippines
300,000 km2 / 95,856,000 pp
Construction Sector Import: 58 [kg/capita]

Singapore
724 km2 / 5,274,700 pp
Construction Sector Import: 1,201 [kg/capita]

Thailand
513,120 km2 / 64,076,000 pp
Construction Sector Import: -49 [kg/capita]

Vietnam
331,210 km2 / 89,316,000 pp
Construction Sector Import: 264 [kg/capita]

Source: ASEAN
Map: Felix Heisel / Marta Wisniewska
ETH Zurich / FCL Singapore

Construction Material Import % in relation to Southeast Asian average

We fundamentally believe in the idea that the city must be a place of production. Innovation has to be driven from the idea of reinventing our industries in smaller, even decentralized units, which do not harm the environment or our health. The knowledge implementation and resulting production of a newly developed bamboo composite material as reinforcement systems in structural concrete could be exactly such an innovative key industry for developing urban territories. It could function as the start of a newly defined industrialization age, whereby questions of social equity, sustainability, environmental protection, education and decentralization could play major roles.

WHAT?

Bamboo Reinforcement Material Sample

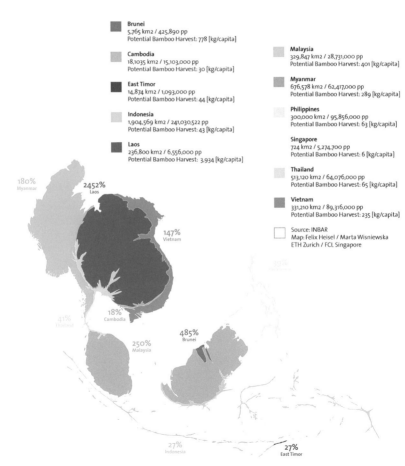

Brunei
5,765 km2 / 425,890 pp
Potential Bamboo Harvest: 778 [kg/capita]

Cambodia
18,1035 km2 / 15,103,000 pp
Potential Bamboo Harvest: 30 [kg/capita]

East Timor
14,874 km2 / 1,093,000 pp
Potential Bamboo Harvest: 44 [kg/capita]

Indonesia
1,904,569 km2 / 241,030,522 pp
Potential Bamboo Harvest: 43 [kg/capita]

Laos
236,800 km2 / 6,556,000 pp
Potential Bamboo Harvest: 3,934 [kg/capita]

Malaysia
329,847 km2 / 28,731,000 pp
Potential Bamboo Harvest: 401 [kg/capita]

Myanmar
676,578 km2 / 62,417,000 pp
Potential Bamboo Harvest: 289 [kg/capita]

Philippines
300,000 km2 / 95,856,000 pp
Potential Bamboo Harvest: 63 [kg/capita]

Singapore
724 km2 / 5,274,700 pp
Potential Bamboo Harvest: 6 [kg/capita]

Thailand
513,120 km2 / 64,076,000 pp
Potential Bamboo Harvest: 65 [kg/capita]

Vietnam
331,210 km2 / 89,316,000 pp
Potential Bamboo Harvest: 235 [kg/capita]

Source: INBAR
Map: Felix Heisel / Marta Wisniewska
ETH Zurich / FCL Singapore

180% Myanmar
2452% Laos
147% Vietnam
41% Thailand
18% Cambodia
250% Malaysia
485% Brunei
27% Indonesia
27% East Timor

Potential Bamboo Harvest % in relation to Southeast Asian average

Education Research Center (ERC), NUS
A Sustainable Learning Complex, Singapore

View from town green

Office W Architects, Singapore
Design period Mid 2007 - Dec 2009
Construction/Realisation period
Dec 2009 - June 2011
Concept by Mok Wei Wei
Team Ng Weng Pan, Chan Kwong Ming, Foo Yong Kai, Darren Tee, Darren Tan, Wong Shu Jun, Bank Ekkachan

Located in the heart of a new University Town, the Education Resource Centre (ERC) of the National University of Singapore (NUS) is an expression of Singapore's endeavour to develop collaborative learning communities for students and researchers to work, live and learn in close proximity. The multi-programmed facility, sited at the centre amidst high-rise residential colleges, faces the town green and operates over extended hours. It is a two-storey building with a part lower-first storey and a roof deck. It houses a 200 seat auditorium, study clusters, a learning café and multiple seminar rooms. With the low building height and a large green roof, it seems to double the area of the town green.

Passive design is a key sustainability concept that is integrated within the architectural design of the building. Where possible, spaces are naturally ventilated. This includes the large foyer, lift lobbies, circulation paths throughout the building and many study areas. The large courtyard that punctuates the building plates also generates a cooling effect that draws cool air down through the open courtyard and induces natural breezes flowing through the building and the naturally ventilated circulation paths. The resultant micro-climate also helps to bring down the ambient temperatures of the enclosed rooms. In addition, the porous design of the building with major entrances and facade openings oriented in the North-South direction also ensures that the cross-ventilation within the building is enhanced by the prevailing breezes. Large rain and sun canopies, as well as vertical green screens, are carefully designed to provide additional shading to the ceramic-fritted windows to further minimize heat transmissions through the glass. One of the key architectural features in the project is a series of 8.5m-tall pigmented fair-faced concrete walls that faces the town green.

The porous design ensures cross ventilation

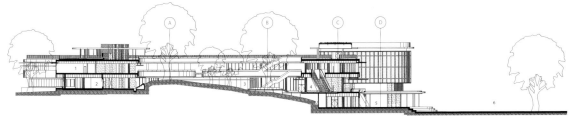

Section 1 seminar room 2 pc commons 3 courtyard 4 foyer 5 learning cafe 6 town green

The building is a **vibrant forum** for students to meet, interact and brainstorm and a place for informal and self-directed learning, group discussion, project work and individual study to support the ease of access to knowledge and technology. It promotes **sustainability,** the use of **information technology,** accessibility of knowledge, continuous adaptation, interactivity and better quality of life. It is therefore an example of a project that "smartens up" our city.

HOW?

The different levels of the topography are negotiated by the fluid geometry of the timber deck.

The open-sided and naturally ventilated foyer has an elevated view to the town green.

Being one of the most prominent parts of the project, the red concrete walls are especially striking when they are lit up at night and seen through the full-height glass glazing of the auditorium.

A large courtyard that punctuates the building plates and enables the magnicent Tembusus to be conserved.

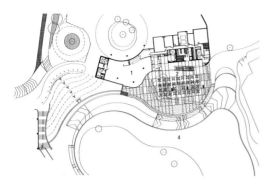

Lower 1st storey
1 indoor learning cafe **2** outdoor learning cafe **3** lift lobby **4** town green

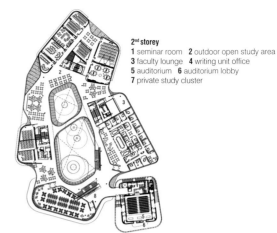

2nd storey
1 seminar room **2** outdoor open study area
3 faculty lounge **4** writing unit office
5 auditorium **6** auditorium lobby
7 private study cluster

WHY?

The 1997 Asian Financial Crisis prompted Singapore to shift from the traditional focus on the manufacturing sector to an innovation driven **knowledge economy.** More knowledge and skill-specific educational institutions were promoted to grow. It is within this context that the National University of Singapore, being the largest and the oldest tertiary educational institution in Singapore, masterplanned the new University. The basis for a lively, intellectual and social students' centre in the heart of the University Town therefore arose.

1st storey **1** pc commons **2** xchange **3** mac commons **4** lift lobby
5 foyer **6** drop-off **7** courtyard

A large plate structure that does not physically compete with the surrounding tall residential colleges.

Walls cast in pigmented concrete act as a feature when viewed from the town green.

With many of the facilities accessible to the students and researchers 24/7, the project encourages **individual learning** by providing a wide variety of study areas generally accessible by the public as well as flexible furniture to be re-configured by the users. With over 1700 power sockets provided, students in the University Town can virtually plug in anywhere they desire to brainstorm for group projects or to work individually. The result is an unprecedented level of flexibility that encourages working and learning beyond classrooms.

The brief required the study areas to be air-conditioned but we recommended a considerable portion of them to be **non-air-conditioned** instead. This not only cuts down energy consumption significantly, but also encourages a healthier way of living and learning. Equipped with hundreds of power sockets for students to 'plug in', the naturally-ventilated Learning Café is well loved by students. Fitted with a Big Ass Fan, a constant gentle breeze flows through the well-ventilated space, making it one of the most popular study areas in the University Town. The project reverses the current trend of enclosing study spaces in fully air-conditioned environments. Instead, it re-introduces interactions and exchanges of knowledge in an **informal, tropical setting** with minimal reliance on our limited natural resources.

WHAT?

The gentle breeze that flows through the space makes the outdoor learning cafe one of the most popular study areas in the University Town.

HOME Units
Low Income Housing for the Elderly, Singapore

The 3 variations of the HOME Unit

Singapore's high densities and high living costs mean that available land for low-income housing is scarce. In response to this situation, a prefabricated mobile compact home for the low-income elderly has been developed. Each unit has a total floor area of 30sqm and comes with fully functioning spaces for everyday life activities. An internal courtyard allows daylight and ventilation.

The HOME Units' 'living' green facades and mobility suggest that they can be placed practically anywhere in the city and will blend in with the urban fabric. The facades are covered with tropical plants to become part of the Singapore government's initiative to green the city with local flora and fauna. Simple hydroponic planting system on the vertical facades will allow growing vegetables and herbs. The idea is to have the authorities' permission to place the units in public areas.

The units come in 3 variations of the main facade design, chosen by the occupant depending on their health, mobility and how and where they prefer to live.

Type 1 comes with a small covered patio, which is suitable for the elderly who prefer to live in an assisted or non-assisted community. This unit type can be placed on the roof top deck of an existing housing block or car park structure or temporary urban open spaces.

Type 2 allows the attachment of a billboard, electronic screen or other interactive medium. The occupant will be able to earn a small income by allowing advertisers to promote their products or services on his/her living unit. This type will be suitable for locations along the high street or at tourist attraction zones.

Type 3 is for the able elderly who would like a living unit that can also perform as a kiosk that gives out tourist information leaflets, sells newspapers, refreshments etc. Similar to Type 2, these units can be located in shopping districts, markets, tourist attraction zones and also public parks.

Office Eleena Jamil Architect, Kuala Lumpur, Malaysia
Location Singapore
Design period July 2012 - September 2012
Construction/Realisation period unbuilt
Concept by Eleena Jamil
Team Yusri Amri Yussoff, Sufiah Muhadzir, Nurnajdah Najib
Coordinator/Client Building Trust International, UK
Website www.ej-architect.com
www.buildingtrustinternational.org
HOME units was shortlisted at the HOME competition (professional category) organized by Building Trust International in August 2012.

HOW?

The HOME Units are designed to **provide housing solutions for a neglected population group,** the low-income elderly, and **provide opportunities of returns in environmental and economic terms** for both the occupant and the city. The units can be used for independent, communal or assisted living and allow the elderly to continue to be active and thrive in society. They are basically work/live pods with natural green facades and parts that can function as an ad space, kiosk or outdoor shelter. The green design aspect of the pods will be a welcoming sight and accepted as part of the urban fabric in the city. While these living units are not meant to be permanent housing for the elderly, they provide a viable solution and choice that is available amongst other possible developments.

LIVING UNIT TYPE 1 with PORTICO

main function: living unit, greening the city

Possible functions: community living (assisted/non-assisted)

Locations: public park, flat roof of a carpark structure, existing housing estate, hospital grounds

Occupant type: one who enjoys living in a community or one who needs living assistance

LIVING UNIT TYPE 2 with INTERACTIVE BOARD

main function: living unit, greening the city

Possible other functions: billboards, electronic media, tourist information interctive kiosk etc.

Locations: high street, public plaza, waterfront

Occupant type: outgoing, active, mobile, able to live independently

LIVING UNIT TYPE 3 with external shelving system (kiosk)

main function: living unit, greening the city

Possible other functions: newspaper kiosk, tourist information kiosk, food kiosk selling fruits, herbs, plants etc.

Locations: high street, public park, marketplace, tourist spots

Occupant type: outgoing, active, mobile, able to live independently

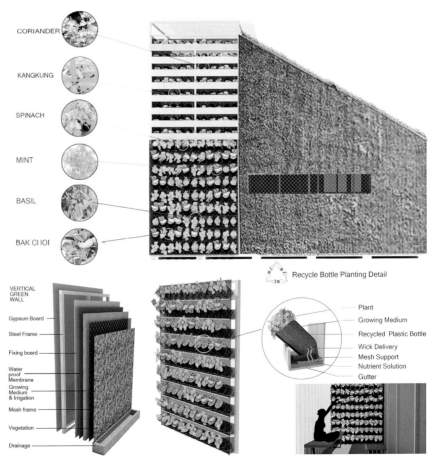

Green wall system of the HOME Unit

WHY?

Singapore has a rapidly ageing population and while many entering the elderly bracket are professionals, managers, executives or administrative workers with good retirement financial plans, there is also **a growing number of the poor and elderly** taking on low-skilled and low-paid employment, with many doing so out of necessity to survive. Because of Singapore's high densities, the **high living costs and the scarce available land** for low-income housing, a large proportion of the local population live in tall tower blocks built by local authorities known as HDB flats and even these are sometimes **beyond the reach of the poor.**

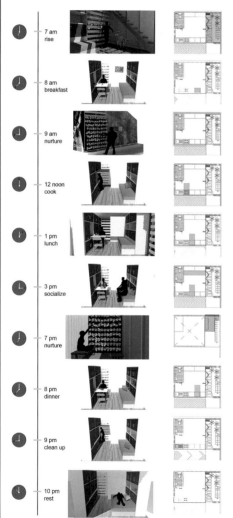

Daily routines within a HOME Unit

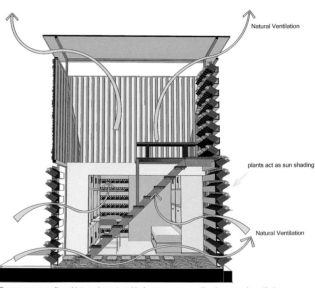

Porous green wall and internal courtyard help encourage cooling by natural ventilation

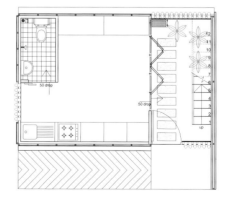

Typical Floor Plan of the HOME Unit

HOME Units can be located on existing roof tops

HOME Units along a shopping street

The HOME Units as kiosks in a park setting

The HOME Units will create **new confidence in the elderly and poor** to go about their everyday lives in the city. Rather than grouping them in a shared accommodation as in most cases, they now have the choice to live as independently as they wish. HOME Units will create **more positive interaction between people from different social and economic backgrounds** and their presence will create an awareness of the current socio-economic situations. It may encourage **a new flexible lifestyle** and economic behavior of combining mobility with work and living.

The green design aspect of the work/live pods will be a welcoming sight and accepted as part of the urban landscape. The occupants must keep their green walls/roofs in thriving conditions in order to maintain their free 'building site'. The challenge will be to convince the local authorities to allow the placement of these HOME Units in public spaces and to **become part of the urban realm** for free or perhaps a small 'license' fee. In exchange the city will gain additional green and billboard surfaces.

WHAT?

Bamboo Bridge
Community-Driven Upgrading of an Urban Informal Settlement, Davao City, The Philippines

The covered arch footbridge over Matina river is a permanent structure made out of bamboo.

Office Arkomjogja, Yogyakarta, Indonesia
Location Matina Crossing Community, Davao City, The Philippines
Design/Research period October 2010 - December 2010
Construction/Realisation period December 2010 - May 2011
Concept by Andrea Fitrianto
Team Joefry Camarista, May Domingo-Price, Natalia Dulcey Garrido, Nassrodin Sapto, Rexan Rainier Cabangal, Aimae Sumilig Juanitas
Building typology 23m span footbridge
Technology & material locally obtained bamboo with borate preservation, bolts and sand-cement grout construction
Partnership Homeless People's Federation Philippines, Inc., Sahabat Bambu (Yogyakarta), ACCA program of the Asian Coalition for Housing Rights (Bangkok), Technical Assistance Movement for People and Environment, Inc.
Website tulaykawayan.blogspot.com

Treated and engineered local bamboo is being used in developing a footbridge in an informal settlement on the riverside in Davao City, the Philippines. Community organizations initiated the project, took loan, and partly funded the project through a savings group. The communities played a central part in the planning, constructing and taking-care of the facility. Situated on lowland in a tropical region, Davao is a fast-growing, medium sized city.

Growing in the tropics and sub-tropics, bamboo is actually a giant grass and among the oldest construction materials known to the human race. For a while modern materials have sidelined bamboo as it is prone to bug plagues, when the stems are improperly harvested. In some places bamboo is associated with poverty or, at worst is considered backward. Resource scarcity and the recent development in bamboo treatment and joinery methods have brought back bamboo's importance as one of today's most promising and sustainable construction materials.

A community leader presenting her group work during the bridge design workshop, February 2010. The workshop was set to challenge the norm of seeing bridges as engineering products. Everyone is a designer. The workshop brought strong local ownership to the project.

Support from the Civil Engineering department / University of Mindanao. The team discusses the foundation design that would deal with lateral the force generated from an arched bridge.

Due to a lack of access, the bridge was designed to be constructed as a "hand made" bridge. After the frames manufactured at the bank, men transfer and install the frames were with the help of bamboo rolls and scaffolding and the bayanihan spirit of community action.

Sand-cement mortar was injected into the bamboo at the main joints. The solid mortar would then transfer the load between the bamboo poles and steel components, bolts and pins.

A layer of reinforced concrete on the floor acts as a shell structure which stabilizes and seamlessly attaches the bridge to the foundation. On a tricycle with a boy from the community is Andrea Fitrianto, a "community architect" from Indonesia who developed the bridge design.

The project used community-owned resources in terms of financing, managing and labor to develop a bamboo bridge as a basic infrastructure, which improves the lives of many. The technologies of treatment and joinery promise 25 years of service with minimum maintenance requirements.

A savings leader welcomes her guests from the women's savings group of Mindanao region. The group is part of the urban poor savings federation Homeless People's Federation Philippines Inc. (hpfpi-pacsii.org) which is affiliated with the Asian Coalition for Housing Rights (achr.net) and Slum/Shack Dwellers International (sdinet.org).

A footbridge can improve access and mobility for household goods as well as people; in safety and comfort, children and the elderly can access city amenities such as schools and hospitals.

HOW?

WHY?

Many Asian cities are situated on riversides or at river mouths. Bamboo groves commonly grow alongside creeks and on riversides, and are known to stabilize soil and to prevent erosion. During crisis situations – concerning energy, economy, resources, and environment – it is time to promote the use of bamboo to help the environment and specifically urban poor families. Southeast Asia is a region of growing economies, resulting in **high urban growth** and increasing informalities. Municipalities are **limited in their capacity to provide their citizens proper housing and infrastructures.** Informal settlements have grown with **a lack of basic services,** such as access to public amenities; some are prone to **risks like flooding.** In poor settlements labor is cheap and social cohesion is present. Pairing appropriate technology and innovative financing can spark local initiatives to grow into concrete action and finally into some meaningful results.

WHAT?

The project illustrates a sustainable model of urban development where alternative technology meets social assets: **community finance,** knowledge, skills, **cohesion,** and **environmental awareness.** The whole process invites wider groups in society, local governments, NGOs, academia, architects and professionals, to interact and contribute, thus to create new networks. Community participation is not a new term in urban development but increasingly demanded. For the bamboo bridge the community formed a savings group prior to the project implementation and run by women in particular. The purpose is to accumulate local resources, not only in monetary terms, but functions as well as a grassroots medium for community management. While women practice savings, men do construction. The **collective work strengthened cohesion** within the community and at the same time **preserved the local wisdom.** Such community-based project also became **a learning space for students, academia, and professionals.** Ultimately, the project stitched the local society together as a whole again.

Warning – Pollen Landing
Shelter Housing, Manila, Philippines

Aerial view of the new productive community

Warning – Pollen Landing sets the fertile ground on which Mandaluyong's relocated slum dwellers can blossom as an autonomous and resilient community. It presents the urban shelter as an empowerment equipped milieu, rather than a simple architectural product, by providing spaces for formal and informal work opportunities as part of different phases of the urban agriculture products' life cycle such as: growing, processing, packaging, marketing and selling.

Harvesting the workforce of the future shelter's beneficiaries by proposing low and simple self-built units, the project insures a low amortisation cost and strengthens the community spirit from the start. Inspired by the hot and humid climate of Manila, it also encourages the participation of kids and seniors in the construction process of a chimney that provides an efficient natural ventilation system made with reused glass bottles, and shading devices in colourful vinyl banners modelled after the vernacular weaving pattern.

The site, Addition Hills, lies in the middle of the municipality of Mandaluyong, Metro Manila.

Office Manon Otto, Montreal, Canada, Student at Lund University Sweden
Location Metro Manila, The Philippines
Design period May 2012
Construction/Realisation period This master student proposal has been developed after a three week field study in Metro Manila as part of the studio "Urban Shelter" coordinated by Johnny Åstrand, architect & director of the department for Housing Development & Management, School of Architecture, Lund University, Sweden. The 7 projects imagined in this studio have been developed with the help of the local NGO TAO Pilipinas inc. and have been presented to the National Housing Authority in order to stimulate the on-going evolution of slum dwellers' relocation programmes in the Philippines.
Concept by Manon Otto
Website www.snoscapes.info

The community's productive roofscape

In front of the community center

HOW?

Detailed typology - Self-Help Housing

At the urban scale: Incremental and careful phasing for development that introduces urban gardening for food security
+ New North-South connection to attach the new neighbourhood to formal work centres
+ Quality public space surrounded by architectural invitation to the surrounding communities
+ Balanced mixed-use and productive urban design.

At the architectural scale: Easy and efficient self-built blocks
+ Super innovative natural ventilation chimney combining three physical phenomenon: the Venturi effect, the solar radiation absorption in thermic mass and the power of concentration of rays from the magnifying glass (reused bottles = abundant local material).

Connecting to the city fabric.

Providing a series of activated spaces.

Integrating the school.

Inviting the surrounding's communities

Maintaining a sponge.

Planning motley clusters.

Creating a threshold, a landmark.

Linking producers and consumers.

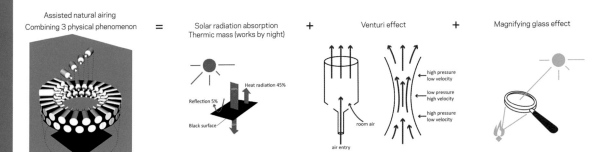

WHY?

In the context of Metro Manila, shelter design for the relocation of informal settlers seems to be achieved as a simple cleaning process: wiping one area after the other, making them shine to reveal their value. Cleaning is a spontaneous act. It does not consider any long-term perspective because the dust that is wiped will fall back anyway. But what if this dust is pollen? What if it lands on the right leaf, on a rich soil that, after some time, could flourish a whole city? 'Warning - Pollen Landing' aims at reversing the paradigm of confining shelters' beneficiaries in gated communities by proposing an inviting and **heterogeneous micro-society**. It defines the shelter as an **empowerment equipped milieu** by providing a variety of **formal and informal work opportunities**. Its sociologically sensitive urban design respects and **encourages local traditions of street life** and informal markets by articulating productive shelters around a market spine.

Inclusive system connecting formal and informal populations around a common urban agriculture project

Duality in formal and informal populations

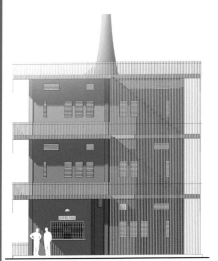

Eastern elevation

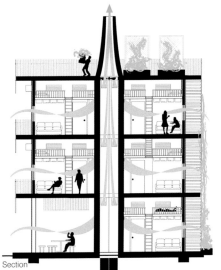

Section

3 levels x 4 units

Firewalls

Flexibility of columns

Coping with weak changing winds

Using a chimney to provide services

Commercial possibility on ground floor

Coping with strong sun

Coping with strong rain

Chimney effect for enhance natural airing

Using a chimney to provide services

WHAT?

At urban scale: Climate-sensitive urban design
+ High density (2300 relocated informal settlers on 1,3ha) achieved with an innovative mix of self-built blocks and building with "amortisaction" periods ("site and services" approach but delivered in multi-storeys)
+ Delivering job opportunities together with shelters so relocated individuals will be able to pay their amortisation by working in a productive community!

At architectural scale: Flexibility in units twining that encourages multigenerational cohabitation + Invitation to affirm sense of community by personalising buildings with shading devices made of reused plastic banners woven like vernacular architecture, a simple and fun task that could involve young and senior citizens in the building process!

Existing street layout

Youth in existing conditions

Playing in a shaded courtyard

San Juan City Plan
Planning Towards a Future of Green Consciousness, Manila, The Philippines

Postcard from the Future: More balanced development on both sides of Annapolis Avenue with vertical developments and elevated tramway.

Office Palafox Associates, Makati City, Metro Manila, Philippines
Location San Juan City, Metro Manila, Philippines
Design Period 2013 - 2023
Construction/Realization Period The CLUP (Comprehensive Land Use Plan) has already passed the legislative readings of the Local Development Council and is currently under implementation
Concept by Felino Palafox, Jr.
Team Karima Patricia Palafox, Urban Planner; Eli Paolo Fresnoza, Urban and Tourism Planner; Pablo Barrios, Architect, BERDE Professional; Geniveve Grace Ranchez, Environmental Planner; Kimleye Ng, Urban Planner; Joseph Peña, Architect-Urban Planner; Albert Tanching, Civil Engineer, San Juan City Resident; Alfred De Guzman, Civil and Sanitary Engineer; Charisma Urcia, Architect, BERDE Professional, San Juan City Resident; Dennis Cruz, Architect; Sean Delos Santos, Planner; Anthony Sarmiento, Architect-Environmental Planner; JP Vallejo, Designer; Jeffrey Caladiao, Designer; Mary Joy De Lara, Research and Development Writer and Researcher; Isabelle Kern - Architect
Website www.palafoxassociates.com

Palafox Associates in partnership with the City Government of San Juan adopted a participatory and inclusionary approach towards the planning of a model green city of excellence through the review and update of its Comprehensive Land Use Plan and Draft Zoning Ordinance (CLUPZO). The City of San Juan is located at the heart of Metro Manila in the Philippines; but despite being the smallest city in terms of area at 594 hectares, the city has had its share of common ills of the urban setting such as traffic due to its narrow roads, proliferation of informal settlers, extreme lack of open spaces, and flooding. The project's general focus was to make the public aware of how the city's urban landscape may be improved through disaster preparedness, improved urban mobility, vertical urbanism, the introduction of mixed-uses, the creation of new and potential urban centers, urban revitalization of the city's waterfront, on-site or nearby relocation of informal settlements along the river, application of traffic management systems, and pedestrianization. Various stakeholders from the business sector, academe, government agencies and social civic groups were consulted and involved in the formulation of the plan. Through workshops and consultations with citizens and benefactors, the Community Specialization and Waterfront Development was chosen as its directive framework strategy towards creating green urbanism.

Existing Annapolis Avenue with its obstructed sidewalks and diagonal parking. Often this area becomes congested with traffic serving as an entry point to the city.

San Juan City skyline. Current Scenario: Structures are low density residential with high-rise condominiums along major thoroughfares and commercial centers.

Postcard from the Future: Rooftop gardens and green walls throughout the city, with predominantly mixed-use developments.

San Juan has accepted the challenge of being the first city to implement proposals that **prioritize people and pedestrians.** The plan will re-shape the city through the application of new urbanism concepts, development principles and global best practices, and **engage its citizens towards a green approach** in city planning. Green consciousness is the answer to pressing issues of climate change and rampant flooding. Priority in the planning process was the recreation of the urban environment through reactive and proactive actions to transform a vulnerable into a sustainable city. Unique was the **integration of disaster risk preparedness** in all macro- and micro scale projects and regulatory policies. Vertical urbanism and mixed-use compact communities were envisioned as alternatives to horizontal sprawling developments, as those areas are preserved for trees and foliage as flood-proofing and mitigation measures. The plan entails **river clean-up programs** and developments to create **leisure and commercial areas along on the riverfront.**

HOW?

San Juan River today, a main water body of the city and one of the major tributaries of the Pasig River System, a major water body of the Metro Manila.

Postcard from the Future: A linear park along the river, connecting the city through water transport, bike paths and pedestrian walkways.

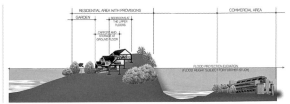

Flood overlay zones require structures to locate living spaces, like bedrooms or living rooms, above the minimum flood protection elevation for the protection of human lives.

WHY?

The city needed to **decrease the risk of climate change impacts.** The planning process has set up guidelines for development and the embodiment of disaster-proof principles; such would prepare its residents for changes in the climate that are not easily controlled. Other issues the project is addressing are aspects of social inclusivity and the **lack of affordable housing,** which has forced the poor and key workers out of the metropolis and travel up to 3 hours to get to work.

Proposed linear parks use 3m easements of all creeks passing through the city with additional provisions for flood protection.

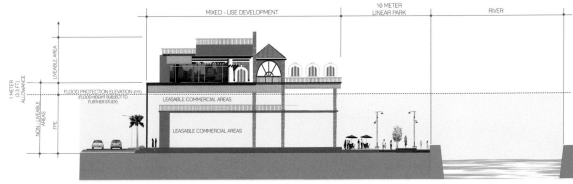

Proposed linear parks using the required 10 meter easements of San Juan River with additional provisions for flood protection.

Existing J. Ruiz Station, an elevated transit station within the city, surrounded by uncomplimentary establishments and dark alleys.

Postcard from the Future: Transit-Oriented Development interconnected with adjacent structures and facilities promotes the sharing of infrastructure and safety of pedestrians.

The plan provoked paradigm shifts, especially regarding the attitudes of top government-level leaders and decision-makers down to residents about environment and climate. It was designed to encourage, building up confidence that such problems can be prepared for. New was the way of implementation: instead of restrictions the plan stimulated with incentives like bonuses for green building designs. It **supported walkable and bikeable alternatives** instead of motorized-transportation and promoted overall better quality of living, and therefore **positive social changes, by design.** Changes and transformations have begun to shape the city such as the fostering and **empowerment of local communities,** a changed outlook of public officials and stakeholders to provide resources and support, and the promotion of public-private partnerships towards long-term solutions.

Ermitaño Creek today, with residents along the creek at risk to flood hazards

Postcard from the Future: Creeks will have landscaped pathways that promote walking and recreational activities.

WHAT?

Smart City Workshop - Manila
March 2013, at Manila Polo Club, Manila, The Philippines

Guided by Dietmar Leyk, Leyk Wollenberg Architects, Berlin; Visiting Professor at The Berlage Center for Advanced Studies in Architecture and Urban Design - TU Delft, under the direction of Joey Yupangco, Dean SDA, Manila

Team Leader Tobby Guggenheimer, Bong Recio, Ed Calma, Dominic Galicia, Alfred Wieneke III

Participating in the workshop were
Joey Yupangco, Christine Darauay, Tina Periquet, Choie Y. Funk, Pearl Robles, Adrian Ace Alfonso, Vince Tan, Carlos Hubilla, Twenty Munoz, Nikki Boncan Buensalido, Mia Quimpo, Asela Domingo, Pinky Poe, Jessica G. Casison, Sudarshan V. Khadka, Laurence Angeles, Larry Carandagn, Marivic Pineda, Conrad Onglao, Mae V. Cu Unijeng, Joseph Gonzalez

An outstanding joining of forces: leading architects from Manila, together with Aedes - International Forum for Contemporary Architecture Berlin, and the Goethe Institut Philippines met for ten days in March 2013 at the Manila Polo Club to rethink the contemporary living conditions in Metro Manila. The workshop was conceived in order to encourage interaction between planners, architects, teachers, and for the future, elected officials and all citizens. The goal of the workshop was to bring together different protagonists engaged in projects and research relating to the transformation of a city. The greater metropolitan area of Manila with a population of 21 million inhabitants became the paradigmatic experimental space for dealing with a Southeast-Asian city in another way.

With regards to the Aedes smart city definition, the debates during the workshop created a very fresh freedom - an inspiring new freedom from masterplans. Consequently the workshop process led to a set of visionary results, which do not necessarily formulate materialization in the city in the first place. Rather than drawing formal models, all ideas circled around a set of smart processes and visionary strategies to sketch out a possible future for Manila by changing the citizen's behaviour. In relation to this the group discussed current demands for the profession of the architect. Not the rejection of form, but to understand how to influence the forces, which lead to the physical manifestation of smart thoughts in Metro Manila.

As in other metropolitan areas in the world, Metro Manila stands for the coexistence of many independently proposed urban development measures. Here the workshop group worked closely together to explore relevant integrated and networked strategies to encourage the rethinking of the city.

Without loosing the broad view for challenges like air pollution, water- and waste management or others, the group focused on three main issues comprising History and Context, Mobility, and the Escolta Neighborhood, the historically meaningful street and architectural heritage of Manila.

March 5th 2013 at Manila Polo Club
Introduction "Smart City" by Dietmar Leyk, Workshop Leader
Presentation of Projects and Participants

March 6th 2013 at Manila Polo Club
Presentation "Social Enterprise Development" GKonomics Design Group, Manila

March 6th - 8th 2013 at Manila Polo Club
Discussion and Work in 3 groups with the topics: History and Context - Mobility - Escolta

March 9th 2013
Presentation and Discussion of Projects at Capellini Showroom
Visit of Escolta Street Market

March 10th - 13th 2013 at Manila Polo Club
Discussion and Work in 3 groups with the topics: History and Context - Mobility - Escolta

March 14th 2013 at Manila Polo Club
Final Presentation with guests from local universities

The ubiquitous presence of walls, physical and non-physical, in regards to safety and social issues, resulted in proposals to instrumentalize these walls in favour of the beginning urban transformation process of Manila.

Mobility, a condition precedent for a modern metropolis, supported by the already existing wide diversity of transportation modalities in Manila, became one of the fundamental drivers for new ideas. Epifanio de los Santos Avenue (EDSA) the longest and the most congested highway in Metro Manila, with an average of 2.34 Million vehicles every day, over 24 kilometres long and passing north-south through 6 of the 17 settlements in Metropolitan Manila, challenged the mobility group to rethink innovative ways to de-congest the movement of vehicles.

Regarding the question of architecture and urban design being a public or rather a private concern in Manila, the decay of the Pasig river embankments played an important role during the debates. Its meaning, strategic location, and its large potential for future interventions through public parks along the river as well as it's meaning for water-transportation became the focus points for many proposals. So do Regenerative Amphibious Floating Terminals (RAFTs), located on land and on water, distributed throughout the entire stretch of the river, become starting devices for a process to populate the embankments with new life.

Rethinking the service of the city instead of its hardware led to new assumptions in regards to technology. The application (app) NetCar for the mobile community complements the informal framework of projects. It offers a platform to share cars in relation to destination and interest groups.

The workshop results deal with intentions and strategies to change the behaviour of people. Both, in regards to the large scale with its implication for Metro Manila, as well as the small scale with its implications for the local context for example the river embankment or Escolta. These projects have the potential to create a cultural awareness in terms of uncovering the potential of the existent, especially the river and the heritage. In combination with fresh ideas about time, performance and choreography of program the projects represent innovative departure points for new urban behaviours in Manila. All proposals comprise ideas about efficiency and capacity, but at the same time they offer fascinating atmospheres and delight. People of Manila are meant to enjoy the experience.

One main success of the workshop is the ongoing collaboration in the workshop group itself. Experts in their own fields, which haven't been working together before, found a common focus point. Due to their interest in an optimistic future for Metro Manila they founded the Manila Smart City Initiative.

Dietmar Leyk May 2013, Berlin

Awarehouse
Upgrading Factories and Employee Housing, Samutprakarn, Bangkok Metropolitan Area, Thailand

Awarehouse consists of 10,000 sqm of warehouse, 320 sqm of office space and 600 sqm of employee housing. Located in a largely industrial sector of Samutprakarn, on the outskirts of Bangkok, Awarehouse immediately distinguishes itself from the ubiquitous windowless, corrugated steel and barrel vaulted neighbouring warehouses with its decidedly alluring pleated form, it's copious perforated steel clerestories, and with its enviable employee housing – clad in hydroponic vegetation – situated proudly in the landscaped foreground of the property. It is the only property of its kind in Bangkok wherein its residents and employees would actually desire to live and work, and be healthier doing so. And, with estimated building costs at just $2,500 per sqm, and with energy costs 40% below those of its neighbours, it is also the one property warehouse owners would most aspire to emulate. Improvements come at approximately only a 5% increase in investment costs with possibly 50% decreases in annual marginal operations costs due to decreases in energy consumption and in employee sick days. Awarehouse transforms in a low-tech modification process, the standard Bangkok warehouse morphology, in order to derive a more environmentally and socially aware paradigm for Southeast Asia's fastest growing and least responsible architectural typology. The Awarehouse housing units have optimal lighting and ventilation, and are situated around a large, two story communal space with horizontal and vertical community gardening.

Office AND Development Co., Ltd., Bangkok
Location Samutprakarn, Bangkok Metropolitan Area, Thailand
Design period 2012
Construction/Realisation period unbuilt
Concept by Taylor Lowe, Ekapob Suksudpaisarn
Team Tarvin Chokdee
Engineer Panusak Kitpanyapan
Collaborators Jarim Weeraboonchai, Lapassanan Buranapatpakorn, Patorn Phoopat, Prompt Udomdech, Varis Niwatsakul, Veerasu Saetae

Awarehouse radically **improves light, ventilation, beauty and workplace standard** of living in a sector of the built environment that otherwise completely neglects such essentials. It lowers artificial lighting and cooling costs; decreases dependence on unhealthy recycled air and enhances passive cooling; increases ventilation; lowers material waste; **promotes urban gardening** to support healthy eating and **improved air quality** in an industrial sector.

Standard Design Poor ventilation, poor natural light, poor convection currents

Warehouse designs are incredibly standardized and are derived by contractors optimizing construction profits, rather than enhancing a working environment. Trade is obsessed with profit maximization and is therefore one of the most significant contributors to migrant exploitation, terrible working conditions and – in a setting like Bangkok where the factory and warehouse labor actually resides at the workplace – lamentable living conditions. As demand for lower priced goods is unlikely to change, and as labor protection is very slow to improve not only in Bangkok but Southeast Asia more broadly, architecture like that embodied in Awarehouse can effect real **change in the quality of life of Thai laborers immediately.**

HOW?

Step 1 Extended roofline to increase sun protection

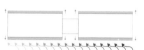

Step 2 Flatten roof and lift up towards the Southeast and North in order to channel natural airflow

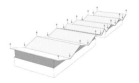

Step 3 Fold roof surface to create 50% North facing and 50% South facing surfaces. North facing skylights will introduce only cooler indirect light

Step 4 Keeping one roofline flat will lower construction costs by reducing scaffolding. The 5 degree slope will further deflect insolation away from overheating the interior

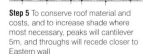

Step 5 To conserve roof material and costs, and to increase shade where most necessary, peaks will cantilever 5m, and throughs will recede closer to Eastern wall

Step 6 Typically windows are distributed evenly, not in any way responsive to the direct Southern light. Particularly with a polded surface, shkylight density should respond to sun intensity

Step 7 Southern facing skylights should be decreased and dispersed, this will save construction costs while also decreasing direct sunlight and therefore energy expenses.

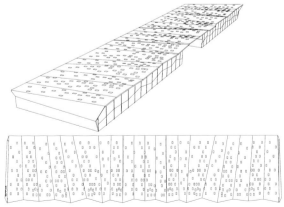

Final Step Skylight density increases on all Northern facing roof surfaces, particularly along the Eastern half where harsh afternoon light will enter the warehouse indirectly. The clerestory is expanded below the longest roof extensions - where shade is maximised - thereby optimising salubrious ventilation and natural cooling.

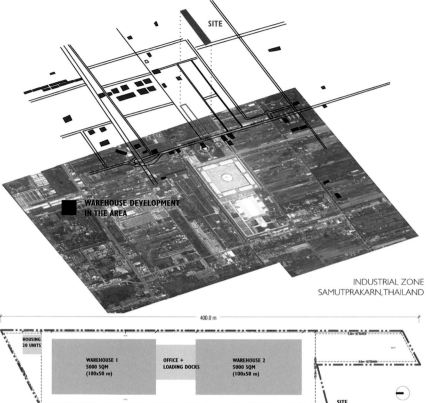

Typical warehouses in Thailand

WHY?

Bangkok is surrounded by warehouses and factories. With 75% of the nation's GDP coming from trade, warehouses and factories are the most commonplace working environment for Thai and immigrant labour. **Warehouses are the fastest growing urban typology** in Thai architecture and, at the same time, **the typology that is most unaware of its social and environmental consequences.** There is, consequently, no sector of the built environment more urgently in need and yet more acutely ignorant of the necessity for improvement than Thai industry.

The EPA (U.S. Environmental Protection Agency) estimates proper insulation and passive design can lower energy expenses by 13-25%. Our design will increase the quality of insulation on those surfaces most impacted by direct sun. The roof and the base of the walls. Clerestories will be maximized to increase cooling air currents through and out of the warehouse.

Awarehouse's worker housing is distributed along a central 2 story landscaped courtyard that offers residents a protected and convivial community space. Most worker housing in Thailand is comprised of punishingly hot galvanized steel shacks concealed at the back of the property.

Central courtyard with community garden and East-West facing hydroponic walls.

Awarehouse provides beautiful and **responsible working and living space at competitive costs** and will help to influence other parties in this problematic system. Socially, Awarehouse **increases visibility of employment practices;** promotes healthier working and living conditions for the disenfranchised working class; **supports urban gardening** and accountability of industrial employment to its labour force both in the work place and at home. Physically, we situated the housing for workers in the front of the site, amidst landscaping, with a degree of visibility that would **ensure the management company invests more into housing** and continues to maintain it.

WHAT?

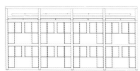

The clerestory below the roof faces North where it channels NE-SW wind currents into the central courtyard. Glazing is optimized where direct sunlight is minimal: along the North facade and along the Southern facade which are continually cast in the shadow of the warehouse.

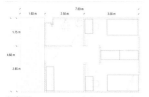
At $150 per sqm, our design economizes on each unit's amenities through a co-housing tactic of sharing restroom and family room facilities among 2 residents. Typically, worker housing provides 1 restroom for as much as 20-40 employees.

Awarehouse, unlike any other warehouse-housing complex in Thailand, exhibits its housing in the front of the property, with ample landscaping, community gardening and hydroponics, to ensure the management's accountability to their workers.

Integrated Park and Streetscape – Chulalongkorn University Centennial Park
Resilient Park for an Overcrowded City, Bangkok, Thailand

The undulating green surface forms various configuration for wide ranges of activities

Office Shma Company Limited, Bangkok, Thailand
Location Bangkok, Thailand
Design period 2012
Construction/Realisation period unbuilt
Concept by Prapan Napawongdee, Yossapon Boonsom
Team Chanon Wangkachonkiat, Nitiporn Sawad, Katavet Sittikit, Suparat Sukrerk, Patchawat Viriyamahattanakul
Website www.shmadesigns.com
www.facebook.com/Shmadesigns

Chulalongkorn University Centennial Park and Streetscape is a part of a mixed-use redevelopment masterplan, which spans over 346,000 sqm in central Bangkok. Given the sizable area of 32,500 sqm, the park is large enough to serve also the existing surrounding neighbourhood within 4 km radius. The streetscape with the area of 39,000 sqm is designed as an extension of the park to link all the plots (in the master plan) back to the park in the middle.

The existing site is full of deteriorating 4 storey shop houses in a rectangular grid segmented with bare road surface with hardly a tree. It was a kind of development that responded well to social needs at the dawn of Thai's modern era when nature was still in abundance. However, with the population that has grown threefold in less than thirty years, this urban renewal master plan is set to cope with new social and environmental shifts.

'Integrated Park and Streetscape' is at the core of the whole master plan. It is set out to be more than just greenery. The park will generate new urban interaction and create a new urban ecosystem that is resilient to increasing environmental challenges.

The park is formed out of an undulating green carpet that stretches over 2 plots of land being divided by a road, combining the park into one single large green surface. The green carpet is raised at the corner as an entrance statement and touches the ground at various points to allow access to the upper side of the park. Various indoor activities are integrated under the carpet including car park, influencing the form to push up higher. This undulating concept is not only responding to the need to camouflage functions below but also offering opportunity to interact with the park creatively such as using the amphitheatre-like space for a play or a band. This strong iconic identity will also induce a memorable park experience for both the existing and upcoming neighbourhood.

Urban fabric of the neighborhood around 1942

MICROCLIMATE : COOL CORRIDOR

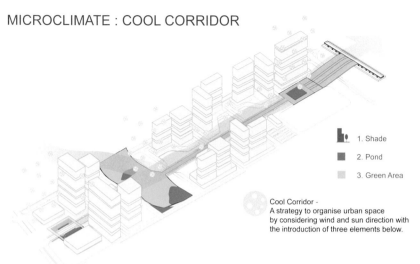

1. Shade
2. Pond
3. Green Area

Cool Corridor -
A strategy to organise urban space by considering wind and sun direction with the introduction of three elements below.

The park is probably the most important public space for Bangkok, but often a gated oasis and not integrated into the urban fabric. Responsive to natural environment phenomena, **the park should become borderless** whether it is out-or indoor or public or private property and a common platform where urban life can be converged and evolved. 'Integrated Park' includes public utilities, recreation, education and commercial activities, relevant to current urban lifestyles. Park and streetscape link to adjacent plots on multiple levels to ensure **seamless connection and accessibility.** The park and streetscape are designed to cope with environmental challenges. Hard pavement is kept to the minimum while **the use of permeable surfaces will allow recharging of ground water.** Storm water management is regulated through landscape elements rather than a traditional closed pipe and sump pit system. **The natural swale, rain garden, and sunken permeable plaza are open to the ground and help to slow down run-off velocity and have water retention capabilities to prevent flooding.**

HOW?

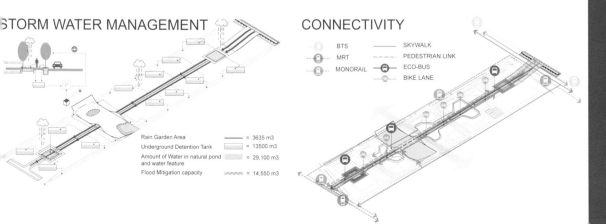

STORM WATER MANAGEMENT

Rain Garden Area = 3635 m3
Underground Detention Tank = 13500 m3
Amount of Water in natural pond and water feature = 29,100 m3
Flood Mitigation capacity = 14,550 m3

CONNECTIVITY

- BTS
- MRT
- MONORAIL
- SKYWALK
- PEDESTRIAN LINK
- ECO-BUS
- BIKE LANE

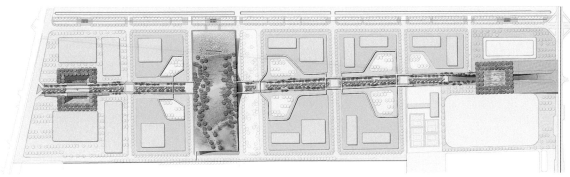
Master plan

WHY?

A) Public space is the common ground for people to share their experience, generate social dynamic/cohesion, a place for recreation, for political debate and it can be a symbol of the city. Unfortunately shopping malls increasingly serving a public function and becoming a collective space of the city much more than a park. 'Integrated Park' will reverse that trend. B) **Unplanned urbanization** has resulted in a **car congested, air polluted,** and built up Bangkok. 'Integrated Park' is based on ecological awareness and contributes to desirable urban living condition. C) Bangkok is built on the silt territory of Chaopraya river flood plain and it is now fully developed into solid territory, which prevents water from penetrating through the ground and also blocks storm water from the north from running into the sea. The Park could be the only public space that can help **mitigate flooding by providing a soft retention area.**

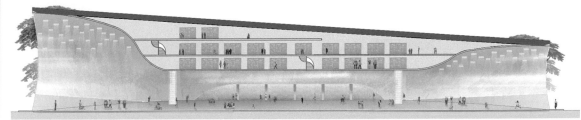
Wisdom Library

Elevation of the park

Permeable streetscape with elevated pedestrian connector

Place for sharing and inspiring

Sunken Plaza connecting urban fabric from both sides of the street

North Plaza with detention capacity during flood

The park's image changes from a static and isolated entity to a more vibrant and connected platform. By **integrating various in- and outdoor functions,** the park can be used 24 hours. **Urban residents prefer the park to shopping malls** and become physically and mentally more fit. Social interaction of people from different backgrounds increases and **leads to a more cohesive society.** Urban residents become more aware of urban ecology and how design is resilient to environmental challenges such as flooding and an increased carbon footprint.

Overhanging 'Wisdom Library' provides shade for public plaza below

WHAT?

Klong Toey Community Lantern
Public Spaces Built with Students and the Community, Bangkok, Thailand

The area becomes full of live and activity as soon as school is over for the day.

Klong Toey is currently the largest and oldest area of informal dwellings in Bangkok. More than 140,000 people are estimated to live here, most in sub-standard houses with few or no tenure rights or support from the government. The area has great social challenges mostly due to the lack of public services like healthcare, affordable education, sanitation and electricity. An extensive drug problem greatly affects the social climate followed by high unemployment rates, violence and crime.

In addition to the main function as a football court and a public playground, the project will work as a tool for the community to tackle some of the social issues in the area. A crucial factor in the continuation of the project is that it will be part of a long-term strategy. This project has to be considered as a small contribution to development on a larger scale, which might lead to positive change. With the connection established both in the local community and a professional network in Thailand, the project has greater chances of having social sustainability. The year-long preparation period allowed the team to design and build the structure in as little as three weeks. The project team got involved with the community through interviews, workshops and public meetings. The design includes several features lacking in the area including new hoops for basketball, a stage for performances or public meetings, walls for climbing and seating.

The main construction's simplicity, repetitive logic and durability enables the local inhabitants to make adaptations that fit with their changing needs without endangering the project's structural strength or the general usability of the playground. This way the project runs in parallel with the ever-changing surroundings and fits with the idea that the project could be part of a larger call for a more sustainable development in the Klong Toey area.

Office TYIN tegnestue, Trondheim, Norway
Location Bangkok, Thailand
Research Period 2010
Construction/Realisation period February 2011 - March 2011
Client Klong Toey Community
Cost 4,500 EUR
Area 91 m²
Sponsors LINK Arkitektur, RATIO Arkitekter AS
Concept by Andreas G. Gjertsen, Yashar Hanstad
Team Architects: Kasama Yamtree, Jeanne-Francoise Fischer, Karoline Markus, Madeleine Johander, Paul la Tourelle, Nadia Miller, Wijitbusaba Marome
Students: Natthanan Yeesunsri, Sarinee Kantana, Nuntiwatt Chomkhamsingha, Nantawan Tongwat, Supojanee Khlib-ngern, Nattaporn Seekongplee, Sarin Synchaisuksawat, Nuchanart Klinjan, Panyada Sornsaree, Porawit Jitjuewong, Amornrat Theapun, Ponjanat Ubolchay, Yaowalak Chanthamas, Boosarin khiawpairee, Praopanitnan Chaiyasang, Kritsana Srichoo, Mario Vahos, Carla Carvalho, Inís Correia, Sarah Louati, Pola Buske, Tabea Daeuwel, Johannes Drechsler, Lisa Gothling, Alessa Hansen, Albert Hermann, Karl Naraghi, Alexander Neumer, Nandini Oehlmann, Fabian Wolf
Website www.tyinarchitects.com

A before-picture of the football court. A lot of work to be done in only 3 weeks.

The structure gives a new character to the street, creating a defined but permeable boundary.

Part of the team, a local architect asks the community children what their new football court should be called.

Parents can now keep an eye on their children while they are playing.

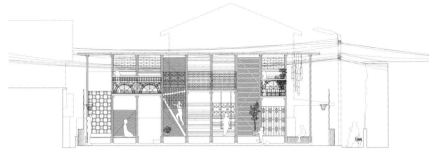

The Klong Toey Community Lantern is a **strategic intervention** that will be part of a positive development on a larger scale. Bangkok struggles with an increasing informal population and very little knowledge, skill and political will to change the current situation. By **connecting professionals, students and the local population in the whole process of planning and building,** we encouraged enthusiasm and exchange of knowledge in urban development processes. The structure in itself is a tool for the community to **create a constructive dialogue between different stakeholders** in this area. It shows that there is interest, initiative and courage hidden beneath the surface of the community.

Urban space is a foreign idea in many of the informal settlements in Thailand. This is not a unique phenomenon, and can be seen in various forms all across the globe. **Lack of ability and will from the local governments** coupled with an undercurrent of **apathy in many of the communities** further weakens the possibilities for urban spaces. By creating structures in the community, both physical and social, and through the support of social interaction, the few open spaces that are available will be protected.

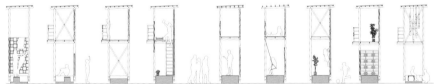
Sections

The Klong Toey Community Lantern is built with a solid foundation and a simple, sturdy main construction. This helps to structure the informal planning practices that go on in the community, and gives the project a durability that isn't connected directly to the physical elements of the building. Before the initiative, the site was used mainly for ballgames and rough play. By introducing more seating, some semi-sheltered places and strengthening some of the unsafe structures that frame the site, **the project encourages a more varied and socially soft atmosphere.** Over time, the project will, and should, **change with the ever shifting needs of the community,** without compromising the main function of the site: **open urban space** and **availability for all of the local population.**

HOW?

WHY?

WHAT?

U-Silk City

Urban Design of 13 Residential High-Rise Towers with Mixed-Use Podium, Hanoi, Vietnam

Side view with central pass through

Office Franken Architekten, Frankfurt am Main, Germany
Location Hanoi, Vietnam
Design/Research period 2008 - 2010
Construction/Realisation period December 2008 - 2014
Concept by Bernhard Franken
Team Christoph Cellarius, Lotte Cellarius, David Berens, Sarah Bolius, René Böttcher, Minh Chu, Eduardo Costa, Hoang Dang, Bui Tien Dung, Thien Duong, Alexander Glaser, Tuan Anh Ha, Hai Khuong, Phuoc Ngo, Hung Nguyen, Thanh Nguyen, Hanno Stehling, Isabel Strelow, Tuan Tong, Thuan Tran
Client Song Da Thang Long
Partner Arup Vietnam Ltd., Space Engineering Co, Transform
Site area 9.2 ha
Dimension Gross floor area 720,000 m²
Website www.franken-architekten.de

This 12-hectare urban expansion in the south of the Vietnamese capital Hanoi is one of Vietnams largest projects today and comprises of thirteen similar yet distinguished residential landmarks for the upper middle class. A public five-storey podium connects the towers and provides retail, food courts, health clubs, kid's play zones, other community spaces and parking for estimated 11,000 tenants. The podium's rooftop offers greenery for leisure activities.

The project had fixed outlines for the towers through an existing masterplan and the challenge was to optimize the design f. e. in terms of building depths, diversity of floorplans and sustainable aspects. The objective was the design of a broad spectrum of apartment types, which is more interesting for buyers in order to match their family size and lifestyle habits. At the same time, the revised floorplans create a vivid displacement of recesses along the façade, the "dancing voids", as the individual facade patterns are generated through recombination and mirroring of the requested unit types. Rather than arbitrary arrangements, the void's displacements follow a rhythm. Each of the residential high-rises has an individual rhythmic structure. The thirteen individual buildings seem to be united in one symphony.

The voids, dividing the building depths, produce more light for the space inside and an excellent microclimate. At the bottom level of each void is a private sky garden that adds value to the individual apartments, and gives the towers a unique appearance.

The project connects to the local history. Silk is a product that dominates Vietnam's history and society. 'U-Silk City' is inspired by the façade texture derived from the computer algorithm, which surrounds the building like a robe and is made of anodized aluminium panels, irregular in their rhythm and oscillate in gentle colours.

Void distribution

Street view

Podium roof top garden

The voids create a new typological approach to construction of multi-storey residential buildings. Instead of a standard floorplan and façade the design offers more variety. It **lifts the landscape from the horizontal to the vertical** by referring to architectural utopias of the vertical city. It includes sky gardens, providing shade, oxygen and evaporation to **improve the microclimate and leisure quality** for the inhabitants. In order to **maximize natural lighting,** the various apartment types have customized recesses tuned to best fit into their room program; the voids serve as **sun protection.** Semi-public spaces are provided in the 1.000-meter long podium for inhabitants and neighbourhood. The podium's **green rooftop** hosts additional leisure functions, like garden and pool areas.

HOW?

Green void as sky garden and "silk" facade

Roof tops

WHY?

Since the introduction of 'Đổi mới' in 1986, a policy to liberalise Vietnams economy, its economy grows at two-digit rates, people's living status has improved, and there is **significant need of advanced infrastructure and living space.** The socialist planning culture lead to top-down urban planning and monotonous and anonymous mono-functional building blocks. This project as a non-standard and multi-use approach creates a more ambiguous and multifaceted urban fabric as a vertical city reminiscent of the horizontal traditional Vietnamese towns.

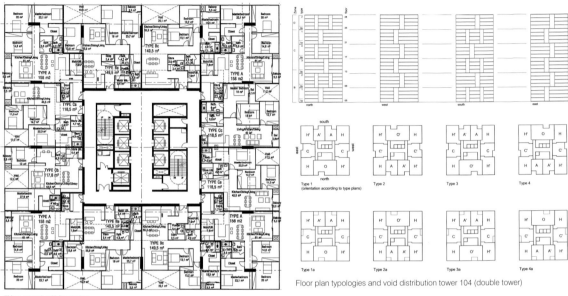

"Non standard" floor plan

Floor plan typologies and void distribution tower 104 (double tower)

Masterplan of the thirteen highrise towers with the podium

Aerial view

The architectural schools in Vietnam are derived from the originally western classical modernist planning culture taught in the socialist block, an engineering driven approach to urban design starting with infrastructure and creating massive monostructural housing blocks. The U-Silk City approach was subversive in a way that it literally deconstructs the original western modernist design by introducing voids. This encourages the Vietnamese architectural discourse to **emancipate from the classical western role model and connect it to the local climate, living style and tradition** and at the same time open it to the global state of the art urban planning discourse. This mixture cannot only create a more adequate Vietnamese architecture but can be a contribution to the global discourse as well.

WHAT?

Construction site of first three towers

Green void as sky garden and "silk" facade

Ho Chi Minh City University of Architecture
New Campus for an Architecture University, Ho Chi Minh City, Vietnam

Designed to blend into the indigenous landscape

Office CAt (Coelacanth and Associates Tokyo), Tokyo, Japan
Location Ho Chi Minh City, Vietnam
Design period 2005 -
Construction/Realisation period currently on hold
Concept by Kazuhiro Kojima/CAt
Team Architects: Daisuke Sanuki, Vo Trong Nghia, Kojima Lab Tokyo University of Science
Landscape Architects: Hiroki Yamashita/Earth Planning and Work, Kayoko Sakashita/Chuoarch
Structural Engineers: Masato Araya/ Structural Design Office OAK
Mechanical Engineers: Saburo Takama/ Scientific Air Conditioning Institute
Client Ho Chi Minh City University of Architecture
Building footprint 42,776 m²
Total floor area 117,008 m²
Storeys 1 - 4 storeys
Structure Reinforced concrete, partly steel frame and precast concrete
Project stage International competition 1st prize 2006
Awards
Asia-Pacific Holcim Award, 2008 (silver)
Global Holcim Award 2009 (silver)
Photo Koji Kobayashi, Sadao Hotta, CAt

This new campus for the Ho Chi Minh University of Architecture is situated in mangrove swamps at the edge of the Mekong Delta in the suburbs of Ho Chi Minh City in South-Vietnam. This university is a National University and the number one architectural education institute in Vietnam. The current number of students is 6000 (prospectively 8000). The campus includes dormitories for 2000 students and sports facilities. Space limitations in the city centre have prompted the decision to relocate the university to this new campus. In 2006, we won the international competition with a solution to the question of how space and education might coexist. We suggested a plan that simultaneously establishes both connectivity and separation for the 6 departments; architecture, city planning, civil engineering, construction, industrial design and fashion.

The project is developed through the integration of the site's water and the mangrove forest and through our concept, FLUID DIRECTION. So called FLUIDS, such as human activity, wind, sunlight and water, are simulated and fed back into the design process. Considering these flows is a necessity for any environmentally and socially sustainable project and, furthermore, liberates architecture from dependence on machines and installations. The arrangement of the buildings, low-rise with a large footprint, was determined by means of an analysis of wind, human activity and sun path. The aim of the design is not to set a large project against the landscape but rather to allow it to blend into the site.

| Go Cong River | Ring Road | Class Rooms | Library | Conference Room |

HOW?

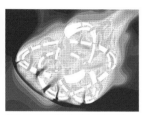

CFD (computational fluid dynamics) analyses were used to create a wind fluid design, eliminating the need for a mechanical air conditioning system in the campus.

By **introducing local materials** into construction, local workers adopt new working methods whilst material transport is minimal, resulting in a more sustainable construction culture. A strong squall that pours down once every day is used for collecting grey water supplies and for cooling through vaporization. The **mangrove forest is restored** with newly raised trees and plants, which grow rapidly in this region, **providing** not only **an impressive visual landscape** but also **shade for the whole site.** "Hardware" costs are low due to minimal land reclamation, use of local materials and low operation and maintenance costs. Thus the **project's budget can be invested in the "software", such as teaching staff and scholarships.**

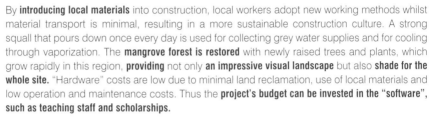

Student activity fluid design: Red represents the flow of people. White indicates furniture (and motorcycle distribution) and other features that invite different kinds of activity. Design is not initiated from rooms and walls, but rather from the arrangement of furniture and the observation of activity.

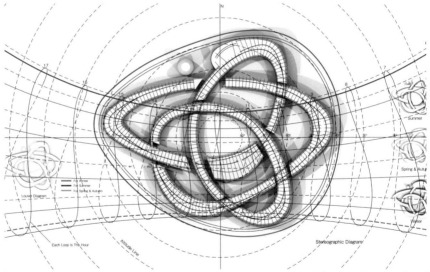

Sun path analysis: Indication of the direction of sunlight in each season. Using computer modeling, the angles and positions of shaded and sunny areas are carefully examined. The building is designed so that direct rays will not penetrate the interior.

Blending into the local landscape

Site plan: The new Ho Chi Minh University campus, located at the edge of the Mekong Delta, integrates the natural landscape of the 40-hectare site.

The Mekong Delta

WHY?

The preservation of the Mekong Delta is of great importance considering the food supply this extremely abundant land provides for the region. Instead of implementing river bank protection work, our idea was to preserve as far as possible the existing scenery, which is covered by water in the rainy season, and to enclose the main campus with a ring road that is higher than the surrounding area. In this project the site is not levelled, as in most new developments, instead it is kept, as much as possible, in its current condition. We have named this strategy of preserving the Mekong Delta by slipping unobtrusively into place "cultivation", and we think it has the potential of influencing planning in the whole region.

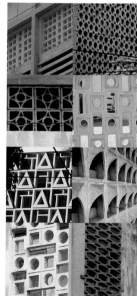
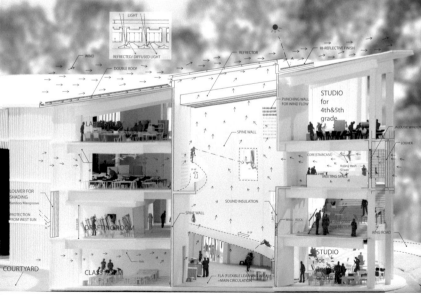

Perforated cement façade materials, seen throughout Ho Chi Minh City, combine ventilation, sunshade and security.

Section Model: The classrooms are organized around a 4-story FLA (Flexible Learning Area), which acts as a part of the main circulation space. The FLA is an informal area for both learning and socializing. Ceiling reflectors deliver indirect daylight by diffusing the light entering through skylights. Taking advantage of the chimney effect provides ventilation.

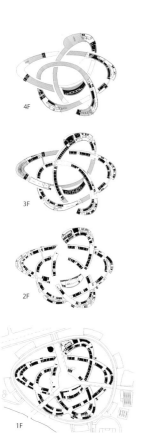

4F

3F

2F

1F

The planning has been determined by defining "Black" and "White" spaces. These spaces differ fundamentally in the way they are used. "Black Spaces" have a one-to-one correspondence between space and usage. For example, in a house this would be a toilet or a closed kitchen. In a school, this would be science laboratories or storerooms. "White Spaces" are not restricted to a single usage. They are fluid spaces that can be adapted to various usages to suit different occasions and needs. The inner side and outer side of the spine wall have differing characteristics. In the inner side are many "Black Spaces" such as classrooms and studios. The outer side accommodates the "White Space", a continuous open space for fluid circulation including open studios, large stairways, meeting spaces, etc.

Junction spaces are designed to allow a breeze to pass through. At the same time, they are the hubs of each faculty, providing space for jury reviews, events and student hangouts.

Junction spaces are designed to allow a breeze to pass through. At the same time, they are the hubs of each faculty, providing space for jury reviews, events and student hangouts.

The building facade is made of bamboo and mangrove, both materials that are produced locally. The facade's porosity allows wind to flow through the building.

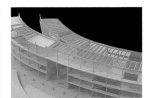

Structural Wall: The main structure is a reinforced concrete post-and-beam structure with brick walls, which is a conventional construction method in Vietnam.

Following the geometry: The "Spine Wall" is integrated into the entire, vast main campus.

Studio zone: Outer side of the spine wall

The "Spine Wall" is arranged so as to link together all of the buildings of the main campus including the library, which is the heart of the campus.

At the crossings of the curves of this project, where different faculties meet, junction spaces are located and function as semi-outdoor spaces, protected from direct sunlight and rain. "White Space" (fluid space in which various usages are possible) **encourages various spontaneous activities like informal learning, socialising**, etc. After analysis, we propose to increase classroom utilization by smart time management and by the additional use of "studios" in the "white space". The result is a more compact and efficiently used building. The idea of **passive building, active people** is used as much as possible. We proposed a campus that would not rely on overall air conditioning and indoor spaces are modified by simple manually-controlled mechanisms. Despite high temperature, humidity and heavy rain a comfortable environment is created.

WHAT?

D7 & D6 Building
Rethinking Urbanism and the Office Prototype, Kuala Lumpur, Malaysia

D6 building - landscaped & pedestrianized street

D7 building - view from Jalan Sentul street

D6 was conceived as one half of a gateway flanking the district Jalan Sentul. Together with D7, D6 heralds in a new phase of urban renewal in Kuala Lumpur. It is rooted in modern architectural tradition where elements of enclosure, structure and mass form the basis of its architectural language. D6 is a simple box, composed as part of a larger ensemble that marks the urban rejuvenation. While the mass is simple, its form is complex in the way the building appears from different angles and at different moments in the day. The complexity is derived from its double-layered glass and mesh skin. The mesh shelters the building from heat and glare and is an invitation for plants and creepers to coexist with architecture. D6 explores the relationship between light and form. The alternating public galleries of the interiors give a tempered experience in lighting. Light falls from a sky lit roof and descends gradually through the chasms of voids. The central theme is conceived to provide offices with a landscape component. The landscape gallery of the 3rd floor allows for visual relief and an environmental surprise. D6 also enjoys a rear water garden. This landscaped strip functions as a quiet 'cooling-off' space with commercial spaces alongside.

D7 is a plastic, malleable, meandering form, which reflects RT+Q's idea that architecture is in essence a sculptural 3D urban construct. The office building around a courtyard has an unusual communal, social and residential character. The courtyard comprises gardens and communal spaces. It also provides for an abundance of natural light, shade and ventilation amidst the tropical weather. The façade is bold, flat and its metallic character contrasts with the prevailing context. The central courtyard is the most important space, designed for social and commercial integration. Outdoor seating, landscaping, patios, transparent elevators and sky bridges are created for 'residents' to enjoy this unique urban experience.

Office RT+Q Architects, Singapore
Location Kuala Lumpur, Malaysia
Design period
D7 2007 - 2008
D6 2007 - 2009
Construction/Realisation period
D7 2008 - 2010
D6 2009 - 2011
Concept by Rene Tan
Team TK Quek, Eddie Gan, Joanne Goh
Website www.rtnq.com

D6 building - view from inside store

D6 building - work environment

D7 building - stores with landscaped views

As a pair of office buildings, D6 and D7 provide users with **a new type of working environment.** They are designed to be close to elements of nature – a provision which a typical office tower lacks. **The ability for the occupants to interact with the natural environment makes them special – and unusual.** Unlike typical high-rise glass-box office buildings, D6 and D7 are low-lying and sensitive to the street scale. They are part of a large urban revival masterplan and form the beginnings of its realization. A link-bridge is just being built to connect D6 and D7, providing easy and safe access between them without having to descend to cross the road.

The commercial spaces filling up with design shops will make Sentul East a vibrant 'design centre' of Kuala Lumpur. **As a pair, D6 and D7 form the urban gateway** and offer a new image of arrival and place for this newly re-energized part of the city. These buildings also aim to offer easy vehicular access, convenient pedestrian mobility – and a different working and living environment.

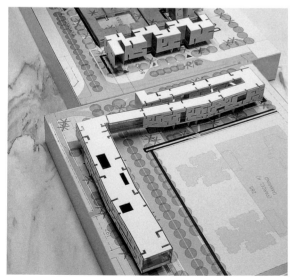
Model of D6 and D7 buildings

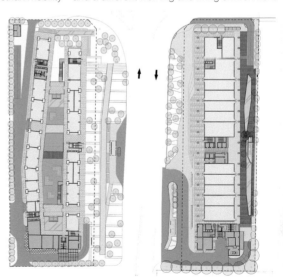
D7 & D6 buildings - ground floor plan

HOW?

D7 building - steps & vertical landscape

New bridge linking D6 and D7 buildings

WHY?

D6 and D7 provide an alternative for Kuala Lumpur as new means of design, lifestyle and work through their non-traditional, non-shop house working environment. This typological variation will inspire a new work order where small Soho type businesses will flourish, especially in this e-intelligent era. They are different from the chaos and clutter that prevent urban growth and rejuvenation. And as Sentul is 'rough' in character, the buildings had to be designed with sensitivity toward maintenance and endurance issues. Maintenance in a tropical climate where the air is humid and damp due to torrential rain and heat, has always been a challenge, shortage of labor also contributes to the general neglect and decay of buildings.

D7 building - central forum in shade

D7 building - skybridges to facilitate access and egress

D6 building - detail of driveway steps

D6 building - light filled alternating corridor

D6 building - studies of opening at steel mesh

The buildings **encourage a different work routine where work can occur both indoors and outdoors.** For example, the 'forum' of D7 allows **social interaction between users and visitors to take place outdoors** in the shade and comfort. It becomes **a new 'hang-out' space.** The advantage is a healthier working environment, which induces productivity. It is 'working in a garden' and with the design shops on the ground level, it provides a new lifestyle to users and visitors – in a 'smart' architectural environment where design awareness, working convenience and work productivity co-exist.

D6 building - front view facing Jalan Sentul

D6 building - internal lift lobby

D6 building - coexisting architecture and landscape

WHAT?

Events accompanying the exhibition

Smart Reception

Friday, 7 June 2013

Kontra-GaP Music and Dance from the Philippines

Words of Welcome
Dr. h.c. Kristin Feireiss Founder Aedes
Jejomar C. Binay Vice-President, Republic of the Philippines
Volker Schlegel President Asia-Pacific-Forum Berlin
Franz Xaver Augustin Regional Director, Goethe Institut Southeast Asia
Ulla Giesler Curator Aedes
Vannita Som, Sok Muygech Winners of the workshops in Phnom Penh

Performance "Singing bricks" by **Irwan Ahmett + Tita Salina** Artists, Jakarta

Symposium / Smart Talks

Saturday, 8 June 2013

Panel 1: Smart Behaviour / Social Inclusion
New concepts for a coherent society support the care for different needs and the inclusion of diverse social groups, to create a productive coexistence in our cities.

Contributors
Prapan Napawongdee Shma Company Limited, Bangkok
Ellisa Evawani University of Indonesia, Jakarta
Eva Esposto Lloyd Collective Studio, Phnom Penh
Thomas Auer Transsolar, Stuttgart
Panelists
Budi Pradono Budi Pradono Architects, Jakarta
Elisa Sutanudjaja Rujak Center for Urban Studies, Jakarta
Moderator
Adeline Seidel Stylepark, Frankfurt

Panel 2: Smart Technologies and Materials
Closed loop economies influenced by new material developments can act as a holistic approach to invent new social value chains for developing countries.

Contributors
Dirk E. Hebel and **Felix Heisel** ETH Zurich/Future Cities Laboratory, Singapore
Eike Roswag Ziegert | Roswag | Seiler Architekten, Berlin
Thorsten Klooster TASK Architekten, Berlin
Moderator
Sascha Peters Agency for Material and Technology, Berlin

Panel 3: Smart Housing
New housing developments can lead to increased innovations with regards to inner-city living. How can new ideas for housing create a change in human behaviour?

Contributors
Eleena Jamil Eleena Jamil Architect, Kuala Lumpur
Kazuhiro Kojima and **Kazuko Akamatsu** CAt, Tokyo
Bernhard Franken Franken Architekten, Frankfurt
Moderator
Kristien Ring Architect and Curator, Berlin

Intervention: On Smart Grids by **Gabi Schillig** and students from the University of Applied Sciences, Düsseldorf

Panel 4: Smart Mobility
To improve the inhabitants' mobility, new transport concepts are key factors for space and traffic challenges in our cities.
Contributors
Max Hirsh ETH Zurich/Future Cities Laboratory, Singapore
Hardesh Singh #Better Cities, Kuala Lumpur
Felino A. Palafox, Jr. Palafox Associates, Philippines
Rainer Becker Head of Business Development Asia-Pacific, Daimler Mobility Services GmbH, Ulm
Norbert Feuerstein Doppelmayr Seilbahnen GmbH, Wolfurt
Christian Derix Aedas, London
Moderator
Florian Lennert Innovation Centre for Mobility and Societal Change, InnoZ, Berlin

Sunday, 9 June 2013

Panel 5: Smart City - The Larger Scale
On what scale can we find significant solutions to the challenges of the twenty-first century and how can we elaborate on the shifting relationship between nature, resource and city?
Key Notes
HE Jai Singh Sohan Singapore Ambassador, Berlin
Dirk Sijmons Curator IABR-2014-URBAN BY NATURE, Rotterdam
Contributors
Kelly Shannon Oslo School of Architecture and Design, Oslo
Eric Frijters FABRIC / IABR-2014, Rotterdam
Lars Lerup Rice University, Houston
Panelists
Cosmas Gozali Atelier Cosmas Gozali, Jakarta
Ben Milbourne Bild Architecture/RMIT University, Melbourne
Moderator
Florian Lennert Innovation Centre for Mobility and Societal Change, InnoZ, Berlin

Panel 5 is supported by and jointly organized with IABR

Intervention: by **Irwan Ahmett** artist, Jakarta

Panel 6: Creative Industries as Smart Economy Factors
Creativity as a social value becomes vital for the urban progress, while relying on global investments, jobs and the young generation as the developers of new economic strength.
Contributors
Rattapong Angaksith Creative City Chiang Mai, Chiang Mai
Tita Larasati Bandung Creative City Forum, Bandung
Ulrich Weinberg Hasso-Plattner-Institut, Potsdam
Gerhard Schmitt ETH Zurich/Future Cities Laboratory, Singapore
Ragnar K. Willer Sociologist of Consumption, Berlin
Moderator
Adeline Seidel Stylepark, Frankfurt

Panel 7: Smart Collaborative Heritage Management
Local governance and heritage management practices in Asia and Europe will be examined with a focus on collaborative partnerships amongst city stakeholders.
Contributors
Nils Scheffler Urban Expert, Berlin
Simon R. Molesworth Executive Chairman, International National Trusts Organisation (INTO), Melbourne
Rene Tan RT+Q Architects, Singapore
Moderator
Anupama Sekhar Deputy Director, Cultural Exchange, Asia-Europe Foundation (ASEF), Singapore

Panel 7 is supported by and jointly organized with ASEF.
ASEF's contribution is with the financial support of the European Union.

Concept **Ulla Giesler** Aedes, **Miriam Mlecek** ANCB

Smart City Research Workshop

Overflowing Potential - The Urban Water Challenge in co-operation with Axor Hansgrohe at ANCB

10 June - 16 June 2013

Concept/Realisation
Dietmar Leyk, Miriam Mlecek, ANCB, Berlin

Speakers
Philippe Grohe Axor Hansgrohe, Schiltach
Rene Tan RT+Q Architects, Singapore
Arno Steguweit Water Sommelier, Berlin
Thomas Willemeit GRAFT, Berlin
Jan and Tim Edler realities:united, Berlin
Nick Meeten Huber SE, Berching
Kelly Shannon Oslo School of Architecture, AHO, Oslo
Antje Stokmann Universität Stuttgart, Stuttgart
Thomas Auer Transsolar, Stuttgart
Yossapon Boonsom Shma Company Limited, Bangkok

Work Group Coaches
Budi Pradono Budi Pradono Architects, Jakarta
Thomas Willemeit GRAFT, Berlin

Guest Coaches
Ellisa Evawani University of Indonesia, Jakarta
Elisa Sutanudjaja Rujak Center for Urban Studies, Jakarta

Participants
Professionals, young professionals and masters students from the fields of Design, Architecture and Engineering, amongst them young professionals from Asia participating in the programme and exhibition Smart City: The Next Generation, Focus Southeast Asia.

Smart Screening
Friday, 14 June 2013, 6PM – 10:30PM

Aedes, together with Asian Hot Shots Berlin and Intimate Moments, is presenting:

FLOODING IN THE TIME OF DROUGHT
Director **Sherman Ong**
Singapore, Malaysia | 184 min in 2 parts
Rotterdam International Film Festival 2010 (European Premiere)

DROUGHT - 92 min | Colour | English subtitles
Language: Hindi, Indonesian, Italian, Tagalog, Mandarin, German

FLOOD - 92 min | Colour | English subtitles
Languages: Japanese, Korean, Mandarin, Thai, Indonesian, Malay, Hokkien
Country of production: Singapore, Malaysia

Discussion
In presence of the director **Sherman Ong** and **Teena Lange**, Asian Hot Shots Berlin

Supporters

The project is financed by:

STIFTUNG DEUTSCHE KLASSENLOTTERIE BERLIN

And supported by:

AXOR hansgrohe    ZUMTOBEL   carpetconcept    BUSCH-JAEGER   TRANS SOLAR

(SEC) SINGAPORE-ETH CENTRE 新加坡-ETH 研究中心

With the backing of:

 Auswärtiges Amt   ASIA-EUROPE FOUNDATION   IABR–   Die Asien-Pazifik-Wochen werden unterstützt durch die Stiftung Deutsche Klassenlotterie Berlin   ASIEN-PAZIFIK-WOCHEN BERLIN

We wish to convey our sincerest thanks to all.

Special Thanks

To all exceptional participants - professionals, students and workshop leaders -, who enriched the exhibition, the workshops and the symposium with their wonderful contributions.
To Franz Xaver Augustin, Regional Director, Goethe Institut Southeast Asia, to Rolf Stehle, Director, Goethe Institut Malaysia and Petra Raymond, Director Goethe Institut The Philippines.
To the colleagues in Southeast Asia: Shelby Doyle, Nico Mesterharm, Meta-House in Phnom Penh, Stefanie Irmer for sharing her contacts, Susiana Wiramihardja, Verena Lehmkuhl and Katrin Sohns, Goethe Institut Jakarta, especially Luisa Zaide for the whole organization of the Manila Workshop, and Joey Yupangco for bringing together the group in Manila. In particular to Marco Kusumawijaya, who generously shared with us his wide knowledge, all his contacts and friends all over the world.
To all Southeast Asian Embassies in Berlin, especially to Ms Milagros Antonio-Kropp, Attachée, the Embassy of the Republic of the Philippines, and Ms Pamela Ong, Second Secretary (Political), Embassy of the Republic of Singapore.

Imprint

SMART CITY: THE NEXT GENERATION

A project by Aedes East International Forum for Contemporary Architecture NPO, in collaboration with the Goethe Institutes in Southeast Asia, with four workshops, an exhibition and a symposium, initiated and curated by Ulla Giesler, Aedes

Exhibition

17 May – 4 July, 2013
Aedes am Pfefferberg
Christinenstr. 18-19, 10119 Berlin
aedes@baunetz.de
www.aedes-arc.de

Curator / Project Management **Ulla Giesler**
Exhibition Designer **Christine Meierhofer**

Smart Loops **Ulla Giesler** Concept, **Christine Meierhofer** Design
Smart Grid Düsseldorf University of Applied Sciences - Faculty of Design **Prof. Gabi Schillig / Rouven Dürre, Tabea Faß, Nadine Hofmann, Marie Christine Keppler, Adalbert Kuzia, Marie Märgner, Nadine Nebel, Jonas Schneider, Anna Wibbeke** with **Michael Swottke**

Accounting **Ramona Bautz, Ulla Giesler**
Assistant **Frieda Krentz**
Textediting **Miriam Mlecek, Stephanie Irmer, Joanna Doherty**
Booking/Technical Support **Moritz Nalbach**
Translation **Galina Green**
Proof Reading **Ramona Bautz, Joanna Doherty**
Set Up **Lothar Schnebel, Benedikt Gnadt, Ricardo Lashley, Chris Cockrell**

Catalogue

Publisher **Kristin Feireiss, Hans-Jürgen Commerell**
Editor **Ulla Giesler**
Design **Christine Meierhofer**
Proof Reading **Ramona Bautz, Joanna Doherty**
Production **HillerMedien**
Printing **Medialis**
© 2013, Aedes and the authors
ISBN 978-3-943615-12-8

Workshops

The three workshops are jointly organized with the Goethe Institut
Workshop Leader Düsseldorf **Gabi Schillig**
Workshop Leader Phnom Penh **Shelby Doyle**
Workshop Leader Jakarta **Elisa Sutanudjaja** and **Dietmar Leyk**
Workshop Leaders Manila **Dietmar Leyk** with **Luisa Zaide** and support of **Joey Yupangco**
Team Leaders Manila **Tobby Guggenheimer, Bong Recio, Ed Calma, Alfred Wieneke III, Dominic Galicia**

Symposium

The symposium is organized together with ANCB
The Metropolitan Laboratory
Concept **Ulla Giesler** Aedes, **Miriam Mlecek** ANCB
Organizer **Dunya Bouchi, Joanna Doherty** ANCB
Intern **Silvia Lucchetta**